Younoussa NOMA TALIBI

Assessment of land degradation and ecosystem services

Younoussa NOMA TALIBI

Assessment of land degradation and ecosystem services

in the Fandou Mayaki Terroir (Tillabéri/ Niger)

ScienciaScripts

Cover image: www.ingimage.com

This book is a translation from the original published under ISBN 978-620-6-72757-6.

Publisher:
Sciencia Scripts
is a trademark of
Dodo Books Indian Ocean Ltd. and OmniScriptum S.R.L publishing group

120 High Road, East Finchley, London, N2 9ED, United Kingdom
Str. Armeneasca 28/1, office 1, Chisinau MD-2012, Republic of Moldova, Europe
Printed at: see last page
ISBN: 978-620-8-33226-6

TABLE OF CONTENTS

ACKNOWLEDGEMENTS

At the end of this work, we would first like to thank ALLAH the Almighty for having given us the time, patience and courage to produce this document.

We would also like to express our deep gratitude to all those who have contributed in any way to its success.

*First of all, we would like to express our gratitude and deep appreciation to our thesis supervisor, Professor **BOUZOU MOUSSA Ibrahim**, who, despite his multiple occupations, agreed to supervise this work. Thank you very much Professor for your availability, your scientific rigour and your advice which led to the production of this document.*

Our thanks also go to the members of the jury:

*Professor **WAZIRI MATO Maman**, who agreed to chair the jury for this examination;*

*Dr **BAHARI IBRAHIM Mahamadou** who agreed to examine and assess this work;*

We would also like to thank Dr TAHIROU Souley, who helped us process the climatic data and satellite images, and for the wise advice he gave us. Thank you, Big Brother, for your encouragement and support;

We would also like to thank the teaching staff and the PATs in the Department of Geography for the training they gave us, their advice and encouragement.

We would also like to thank the head of the Fandou Mayaki sector (DJIBO SALAH), who helped us a great deal during the data collection phase, as well as the entire population of the various villages surveyed, and especially the family of Issoufa MOUSSA, who provided me with accommodation during the field phase. I would like to express my gratitude to them, and may Allah reward you and add this to your good deeds.

We would also like to thank the class of Master2 2020-2021 in general and our research team in particular. Thank you for the time shared together, the fruitful exchanges and above all for the patience you have endured throughout this training and research phase.

And to all those whose names are not mentioned here and who have contributed in one way or another, I would like to express my gratitude and thanks.

DEDICATION

*I dedicate this work of study and research first of all to my very dear parents: **NOMA TALIBI** and **RAMATOU HAINIKOYE** for their efforts and sacrifices made throughout my life and from which I remained far away during my studies. I would like to express my gratitude to them for everything they have always done for me.*

May Allah reward you and give you long life, dear parents. AMEEN !!!!!! Then to my late guardians:
ADA BANA BACHIROU MAMAN SEYDOU SAWANI
May Allah have mercy on you and forgive you your sins;

May Allah make this a Work that will lead you to Paradise.

AMEEN !!!!!!!

Beyond that, this memoir is dedicated to all those who are dear to me. May they find here the expression of my

deep gratitude that words cannot describe.

SUMMARY

For several decades now, the Sahel in general and Niger in particular have been experiencing extreme climatic variations. This situation, combined with strong human pressures, has led to a loss of vegetation cover, exposing soils to the effects of erosion and structural and diverse degradation. As a result, natural resources are being degraded and the value of ecosystem services is being lost. The municipality of Tondikandia is not immune to this situation, hence the interest shown in assessing land degradation and ecosystem services, and in proposing development alternatives. To carry out this study, qualitative and quantitative data were collected from the heads of farming households in four villages in the study area, climatic and granulometric data were processed, analysed and interpreted, and satellite images were processed to confirm what was observed in the field.

The results of these analyses show that land degradation in this Terroir is at an advanced stage, which has led to a disruption in the provision of ecosystem services. The factors behind this degradation are linked not only to climatic variations but also to demographic growth, which is putting considerable human pressure on the environment. Faced with this situation, the population has adopted and continues to adopt a number of measures, such as the use of organic manure, chemical fertilisers and household waste. But despite these efforts, they have not had a significant impact on the sustainable management and restoration of already degraded land. Alternatives have therefore been proposed, ranging from good management measures to techniques for restoring land and ecosystem services.

Keywords: *Land degradation; ecosystem services; Tondikandia commune*

INTRODUCTION GENERAL

Land degradation is a phenomenon that reduces or destroys the capacity of soils to be productive. It is one of the major problems for the future of a planet that is becoming increasingly anthropised, particularly in developing countries. Such degradation is inherent, first of all, in the location of land on the very surface of the globe, which puts it in direct contact, on the one hand, with natural atmospheric elements and, on the other, with the various actions associated with the most common human interventions (agriculture, livestock farming, etc.). This degradation is also related to the very nature of the soil. We know that soil is above all a loose material, made up of the juxtaposition of three kinds of elements: skeleton (minerals - plant debris), plasma (clay-humus) and living organisms (roots and telluric mesofauna), elements which, when they interact, ensure the maintenance of the whole and give it a certain stability, (P. BARBANT, 2010).

However, any natural or man-made action on the earth's surface that inhibits plasma production or affects the links between the different phases that make up the soil, encourages land degradation. Soil degradation is all the more marked when the soil is highly vulnerable, when the action of the climate is more incisive, when human pressure is more accentuated or when the action lasts longer. These factors also have an impact on ecosystem services. Indeed, global climate change and the needs generated by population growth are likely to further degrade ecosystems over the coming years, increasingly calling into question current assumptions and practices (L. Jonathan et al. 2009). This has been highlighted by the Millennium Ecosystem Assessment, which found that over the last 50 years, the decline of ecosystems has accelerated and spread at a rate unprecedented in human history. If this degradation is not brought under control, it will jeopardise not only global biodiversity but also the planet's economic activities. According to L. Bruno et al (2015), most biological communities are disappearing under the combined effect of climate change and other anthropogenic pressures, affecting the functionality of ecosystems that are close to collapse.

Niger, a large Sahelo-Saharan country with a surface area of 1,267,000 square kilometres and an estimated population of 21,942,944 in 2019, with natural intercensal growth of 3.7% per year (MINISTERE DU PLAN, 2020) and recurrent climatic variations, is not immune to this situation. Indeed, like most sub-Saharan African countries in the circum-Sahara, Niger is ravaged by a pronounced phenomenon of desertification and land degradation, (S. ADDAM KIARI and I. ABOUBACAR, 2016). This degradation is due to a combination of factors such as climatic variability and demographic pressure, resulting in increased pressure on the environment. This is manifested in the formation and widening of ravines, often on cultivated land; the formation of shifting dunes; the silting up of farmland, watercourses, agricultural production basins and various socio-economic infrastructures (roads, housing, etc.); the salinisation of irrigated farmland; the leaching of nutrients; soil crusting; the reduction in plant cover and the loss of biodiversity (Idem). This also leads to the degradation of ecosystems, resulting in a reduction in the services they provide to populations. The effects of this degradation and its various forms are naturally reflected in the disorganisation of production systems, the decline in rural production (particularly agriculture, livestock farming and forestry), the fall in household income and the persistence of food insecurity, which are felt much more acutely in Sahelian countries.This being said, humanity's future depends on a good understanding of the phenomena of degradation, as well as the methods for rehabilitating land that is still recoverable. Hence the interest of this study, which focuses on the "assessment of land degradation and ecosystem services in the Fandou Mayaki area (Tondikandia/ Tillabéri rural district) and management proposals". This can provide important data both for better management of current development operations and as a reference point for future actions undertaken by human societies. A methodology based on socio-economic surveys, analysis of climatic and granulometric data and processing of satellite images was adopted for this study.

This work is structured in four (4) chapters, the first of which deals with the theoretical and methodological framework and the presentation of the study area, the second addresses the question of assessing the state of land degradation and ecosystem services, the third studies the factors of land degradation and ecosystem services and the fourth deals with the discussion of the results and management proposals.

CHAPTER 1
FRAMEWORK

This first chapter groups together three (3) essential parts of the body of this work, namely the theoretical framework, the practical framework and the presentation of the study area.

1.1. THEORETICAL FRAMEWORK

This first section sets out the context of the study, and then reviews the state of the art of related work.
with this subject and, finally, sets out the issues to be addressed in this study.

1.1.1. Background and justification

For decades, Niger has been facing a continuous deterioration of its environment. The factors behind this degradation are numerous and complex in their interactions. Indeed, factors such as frequent droughts, rapid population growth, accelerated land erosion, dwindling plant cover and the impoverishment of plant and animal biodiversity are combining to produce a worrying desertification (A. MAHAMANE et al. 2009). These frequent droughts and population growth combine in an insidious synergy to produce accelerated, sometimes irreversible, degradation of certain ecosystems. These ecosystems are being overexploited by the population to satisfy their energy and food needs, especially in areas with high human concentrations. These factors also favour a loss of plant and animal biodiversity, leading to a reduction in life alternatives and a decline in the quality of living space.

The western part of Niger, which is part of the Sahel zone, has a semi-arid climate and is particularly subject to intense environmental degradation as a result of climatic variability and human activity. The latter have led to the over-exploitation of natural resources, making them increasingly scarce. This scarcity of resources can be seen in the shrinking or widening of water bodies, the disappearance of plant cover and the crusting or compaction of soils. As the pores of these soils close, they are subject to a lack of infiltration and become more or less conducive to run-off during the rainy season. This encourages rapid drying out. As a result, the ongoing degradation of natural resources in these semi-arid environments, leading to a decline in ecosystem services, makes food security in these areas precarious.

As a result, in the west of Niger in general, and in the Tondikandia commune in particular, these resources are deteriorating, probably as a result of overexploitation, which is increasing not only to meet the needs of the local population, but also to meet the needs of the local economy. the local population and to feed the city of Niamey. The land of Fandou Mayaki, which is the subject of this study, is also not immune to this situation. Furthermore, according to our In our readings, we noted that this commune in general and the terroir in particular had not been the subject of enough studies on this phenomenon. From a scientific point of view, we have the impression that the research carried out on this type of issue in this commune has not produced the expected results. So there is still a lot to be done in terms of rational and sustainable management of natural resources, especially in the western regions of Niger, which are under serious threat from degradation. And to deal with this phenomenon of degradation, it is clear that more needs to be known about it in order to better understand it and find ways of mitigating its effects, which are hampering agricultural and fodder production. For all these reasons, we have chosen to focus on this area, where the degradation process is accelerating and becoming a major concern day by day.

1.1.2. Review of literature

According to P. BARBANT (2008), land degradation is a process that reduces or destroys the capacity of land for agricultural, plant, animal and forest production. It results from human activities or is a natural phenomenon aggravated by the effect of human activities.

According to the UNCCD, it is defined as a reduction or loss in the productivity and biological or economic complexity of non-irrigated cropland, irrigated cropland, rangelands or pastures, forests and woodlands. It corresponds to the reduction in the economic value of ecosystem services and goods derived from the land as a result of human activities or natural biophysical change (ELD initiative, 2019) taken from ELD initiative, (2013).

This phenomenon, which is defined by several authors, is recognised as a major problem for development and a challenge that is both global and local. This is why a number of studies have been devoted to it at both international and national level. To study this phenomenon, and without being too exhaustive, it would be necessary to review a number of scientific documents and articles on the subject. This will enable us to understand the causes of this phenomenon, its consequences and the writings that have dealt with the various measures and techniques for managing and conserving water and soil.

1.1.2.1. Factors in the degradation of land

Land degradation is a phenomenon caused by a number of factors, the most frequently mentioned of which are demographic pressure and climate change.

1.1.2.1.1. Land degradation and demographic pressure

According to a number of studies, land degradation is seen as a phenomenon that stems from ever-increasing population growth, leading to over-exploitation of the natural environment. For example, P. BARBANT (2008), by comparing the total surface area of emerged land, that of arable land and degraded land, shows that land degradation has increased sharply over the last 60 years or so in the world as a result of demographic growth and industrial expansion. To address this issue, he adopted a method that involved collecting, exploiting and structuring basic data; making field observations to determine degradation indicators; and finally using these variables to determine the synthetic land degradation index. The results of this methodology enabled him to draw up firstly a provisional map with delimited and identified polygons on which roads and drainage axes are represented, then a degradation status map on which the various degradation indicators are represented and finally a zoning map of the land characterised by a degradation index, ranging from 1 to 5. According to its findings, land degradation has far more serious immediate social, economic and political consequences in the countries of the South than in the countries of the North. This degradation phenomenon can be summed up as soil erosion, which is one of the consequences of demographic growth. This form of degradation has been tackled through various approaches. In extreme North

Africa, the concentration of populations around the Mediterranean basin over the centuries has also led to the degradation of plant cover, soils, rivers and climate (R. CHEBBANI et al. 1996). This degradation has encouraged the phenomenon of erosion, especially water erosion, which is dangerously affecting the ISSER catchment area (Algeria). By estimating the amount of land affected by this erosion, i.e. 12 million hectares, and the volume of water lost, i.e. 25 million cubic metres, they state that water erosion is a very serious phenomenon in Algeria. To study this phenomenon, experimental data were linked to the hydro-morphological and analytical characteristics of soils, particularly their surface condition, at different spatial scales.

The results of this study showed that this is manifested by the recession of the head of the gully, the hollowing and scouring of the bottom of the gully, and the lateral scouring of the banks by the effect of the mass of water.

As a result of this same demographic pressure, other underlying variables are also considered by researchers to be determinants of degradation.

For E. KOIHO et al (2017), this degradation is also linked to poor management of natural resources. By conducting a survey of sixty (60) pilot farmers in the Centre West region of Burkina Faso, these authors show that material, financial and technical constraints, as well as those linked to the poor organisation of agricultural markets and land tenure insecurity, are the main causes of degradation. determining factors. In their view, the adoption of good practice in sustainable land management and a heightened awareness on the part of stakeholders, combined with consistent and coordinated support, could lead to better management of natural resources and lay the foundations for sustainable economic and social development.

The same country (Burkina Faso) is also witnessing a steady decline in soil fertility, particularly in its organic matter content, which is compromising the sustainability of agroforestry production (B. OUEDRAGO et al.2019). This is also the case in Benin, where I. AMONMIDE et al (2019), conducting a study on the assessment of soil fertility levels in cotton-growing basins, using agronomic analysis methods, show that most of the soils in this zone are degraded due to heavy exploitation of the environment, the continuous increase in cultivated areas resulting in a low organic matter content of less than 1.52% and a low nitrogen content.

Niger, an agro-pastoral country, is also facing environmental degradation linked to the growth of its livestock population. This situation is leading to a growing demand for animal products, and livestock farmers are using natural standing and harvested fodder to meet this demand (A. DAN GOMMA et al.2019). These authors assume that this fodder collection affects peri-urban ecosystems. Using perception surveys combined with measurements of organic matter, forest inventories, and counts of bare areas and gullies on a hectare, they confirm that this practice has harmful effects on the environment because it causes the loss of plant biodiversity, reduces organic matter, impoverishes the soil and degrades the land. This situation is much more acute in the peri-urban area of Niamey (Tillabéry) than in Maradi, due to the high demand for fodder products. M. AUDRIN (2005), studying the infiltrability of soils in a Niger watershed (Wankama), states that Sahelian soils are in danger.

As a result of demographic pressure and climate change, these areas are undergoing a process of unbridled degradation. To carry out this study, field missions were carried out to measure

soil infiltrability, through which maps were produced to spatialise the results over the basin. On the basis of these results, he sees human activity as the causal agent of this phenomenon, as galloping growth means that people are clearing the land and destroying the plant cover, exposing the soil to erosion. He also noted that the percentage of land in the basin under cultivation had risen from 6% in 1950 to 17% in 1975 and 57% in 1992, an increase of more than 50% in less than 50 years. This has led to fields spreading out as far as the village.
Apart from this factor, climate change is also seen as a contributing factor to land degradation.

1.1.2.1.2. Land degradation and climate change

Climate change is defined by the IPCC (2013) as a change in the state of the climate that can be detected (for example, by statistical tests) by changes in the mean and/or variability of its properties and that persists for an extended period, typically decades or longer. Climate change may be due to natural internal processes or external forcings, such as modulations of solar cycles, volcanic eruptions or persistent anthropogenic changes in atmospheric composition or land use.

Article 1 of the 1992 United Nations Framework Convention on Climate Change (UNFCCC) defines climate change as "a change of climate which is attributed directly or indirectly to human activity that alters the composition of the global atmosphere and which is in addition to natural climate variability observed over comparable time periods".

The effects of population pressure and unsustainable land management practices on land degradation and desertification are being exacerbated worldwide by the effects of climate change, which include changing precipitation patterns, increasing frequency and intensity of droughts and floods, rising temperatures and profound ecological changes (UNCCD, 2015).

Through its various threats, it is considered to be the main factor in land degradation in the world in general and in the Sahel in particular.

According to the United Nations Convention to Combat Desertification (UNCCD, 2015), climate variability is a primary driver of land degradation in drylands. Given the high temperatures and low rainfall that most of these areas already experience, these regions are particularly susceptible to damaging interactions between land degradation and climate change.

Indeed, climate-induced changes in air temperature and humidity, combined with reductions in soil organic matter, above- and below-ground biomass and soil fertility caused by degradation, can exacerbate desertification. In addition, increasingly frequent extreme weather events, such as severe droughts or heavy rainfall, are likely to exacerbate wind and water erosion and contribute to further reductions in biomass, and physical and chemical land degradation. In response to this situation, the convention proposes a number of resolutions such as adaptation, the implementation of agroecological systems, livestock grazing and sustainable land management. These enable soil fertility and watering efficiency to be increased in cropping systems, ecosystems to be enhanced, organic matter reserves and soil biodiversity to be restored, and significant quantities of carbon to be stored. This helps to mitigate climate change.

The effects of this climate change can be seen in Tunisia, where water erosion is widespread and is also considered to be one of the main sources of soil deterioration in agricultural areas, both on a local farm scale and on a catchment scale (M. FOURATI et al.2015). These authors carried out a study of the Oued Sidi Salah catchment (Sfax, Tunisia), showing that water erosion is highly prevalent in the area as a result of a rainfall pattern characterised by highly irregular, torrential and violent rainfall. By adopting a cartographic method based on the use of image processing and GIS software, they were able to draw up maps showing the location of the basin, the hydrographic network and the erosion. These maps show that the basin under study is suffering from the problem of recession at the head of ravines, the creation of new waterways often to the benefit of agricultural land, and significant gully erosion concentrated mainly in the upstream part of the basin and even in the crusted, gently sloping terrain. This idea is in line with that of D. ABDOULAYE and N. AMADOU MOCTAR, (2010) who assert that the countries of the Sahelo-Saharan zone of the Sahara circuit have also been confronted for several decades, with the exception of a few years of normal rainfall, with a rainfall deficit.

This situation has led to major disruption of the major ecological complexes, a downward trend in isohyets towards the south and an almost inexorable process of desertification.

The same phenomenon can also be observed in Algeria, which has experienced spectacular water erosion in recent years as a result of fluctuating climatic parameters. The Djelfa region in central Algeria is one of the areas most affected by this phenomenon (M. KOUSSA and T. BOUZIANE, 2018). It is characterised by highly irregular rainfall, strong aridity, steep slopes and low vegetation cover, making it highly vulnerable to erosion. To gain a better understanding of this phenomenon, the mapping method based on the USLE model was applied to determine the rate of erosion by sheet runoff, map the areas at risk of erosion and popularise the results for appropriate decision-making. This method was used to produce thematic maps (slope, topography, hydrographic network map), the results of which show that the soils of the Djelfa region are affected by several factors that favour erosion, namely the degradation of the plant cover due to persistent drought, which increases soil erodibility. These soils are subject to the erosive effect of splash rains. They also indicate that the regions of Masàad and The Hassi bahbeh areas are subject to heavy erosion, with losses ranging from 1.5 to 23 t/ha/year, which is higher than the model normal of 0.31 t/ha/year. This is due to their geomorphological characteristics, the densification of hydrographic networks and the rise in slopes in these areas. The various effects of climate change (recurrent droughts, increased rainfall, violent winds, etc.) are encouraging land degradation, especially in Sahelian countries. In other words, it is the combination of these factors that leads to the most exacerbated land degradation.

1.1.2.1.3. Demographic pressure, climate change and degradation of land and ecosystem services

The degradation of land and ecosystem services may also be due to a combination of these different factors. The result is over-exploitation of the land, often beyond the real capacity of ecosystems, increased vulnerability of the land and a considerable loss of its production potential. This situation can be seen in several countries in the Sahel-Saharan zone.

In Burkina Faso, worsening climatic conditions have led to a massive influx of people from the north to the west, creating new agricultural pioneer fronts (B. OUEDRAGO et al.2019). This situation, combined with unsuitable agricultural production systems, has led to vegetation degradation and increased water erosion in this part of the country. Using modelling based on the integration of the main water erosion factors into a geographic information system, they highlighted the erosive impact of each factor and ranked the areas at risk of water erosion in the study site. The results obtained enabled an erosion map to be drawn up showing land loss values ranging from 0 to 1.57 t/ha/year over an area of 187196.54 ha and from 8.42 to 36.35 t/ha/year covering an area of 362.11 ha. Studies carried out by a land management and resource conservation project in the central plateau of Burkina Faso in June 2012 also show that in the area covered by the project's activities, for example, more than 25% of families left their villages between 1975 and 1985 following droughts that led to the loss of an average of 20t/h of soil per year, with the extension of bare and degraded surfaces accelerating and the water table dropping by one metre per year (same authors).

This phenomenon of degradation has not spared the lowlands. This is the case of the lowlands in the Nakanbé Dem sub-catchment (Burkina Faso). These lowlands, which were once considered uncultivated land for agricultural purposes, are now used and coveted by herders and farmers because of their potential, and are subject to various threats (A. ILBOUDO et al.2020). According to In their survey of lowland farmers, they identified mechanical ploughing, flooding, strong winds, overgrazing, frequent droughts, bush fires and logging as the main factors in the degradation of the lowlands. This degradation is marked by gullying of the lowlands, sand and gravel plating, and a reduction in plant cover and the number of trees around the lowlands. This idea is in line with that of I. BOUZOU MOUSSA and N. DAN LAMSO, (2004) who assert that the physical and socio-economic constraints facing semi-arid regions mean that their soils undergo accelerated degradation, resulting in low productivity.

In addition, as a result of the high levels of human activity and increased rainfall seen in recent years in certain catchment areas, these areas are undergoing changes to their surface area, sometimes reinforcing their hydro-erosion dynamics.

This is the case of the Kori Mountséka (Tahoua) which, after a long period of dysfunction, has become functional again (I. BOUZOU MOUSSA et al.2009). In this study, the method used consisted of carrying out field missions, collecting, analysing and processing rainfall data using statistical software, and satellite images were used to produce water erosion maps. The results of this research showed that this phenomenon can be explained by the fact that this watershed is subject to intense hydro-erosive dynamics linked to the erodibility of the soil and the weakness of the vegetation cover, which are the result of heavy land use and improved rainfall, both of which contribute to increased run-off and erosion. These processes result in the stripping of surface soil horizons and gullying, which in turn contribute to the degradation of the land. This land degradation has a major impact on the ecosystem services provided by natural resources. These are defined by the Milenuim Ecosystem Assessment (2005) quoted by W. CHRISTIAN et al. (2015) as all the benefits that ecosystems provide to humans. For DAILY, (1977) reported by F. RIVES et al. (2016) "Ecosystem services are the conditions and processes by which natural ecosystems, and the species that constitute them, support and satisfy human life. They maintain biodiversity and the production of goods derived from ecosystems, such as fish, borehole water, wood, biomass fuels, natural fibres and many pharmaceutical or industrial products (...). In addition to this production of goods, ecosystem services are the real life support functions, such as purification, recycling and renewal, but also the source of many intangible cultural and aesthetic benefits". These ecosystem services are grouped into four main categories: provisioning services, regulating services, cultural services and self-maintenance services (W. CHRISTIAN et al. 2015).

So, for supply services, we include all production from the soil such as :
- products intended for human consumption, but also for farm animals;

- wood from forests used for energy or construction;

- fibre for the production of textiles (cotton, flax, etc.);

- raw material extracted from the soil to be used as horticultural support (peat, clay) or as building materials (cob, mud bricks);
- the genetic resource derived from the soil's internal biodiversity, which provides biochemical and pharmaceutical products in particular.

The physical support function provided by soils is also included in this category of services: soils represent the physical base on which animals, humans and all infrastructures stand. The importance of these provisioning services provided by soils is obviously considerable (W. CHRISTIAN et al. 2015).

Regulating services include: regulating the quantity and quality of water transfers between the atmosphere, groundwater and watercourses, which act as water reserves for plants and soil organisms; biodiversity, which plays an essential role in all soil biogeochemical processes, particularly those affecting the carbon and nutrient cycles, especially nitrogen. By modifying soil porosity and cohesion, soil organisms also contribute to soil structuring. This influences not only the physical properties of the soil, but also its water content; waste is recycled through the transformation and transfer of important elements in the natural cycle. This recycling is particularly important for phosphorus, an essential nutrient for plant production (W. CHRISTIAN et al.2015).

As far as cultural services are concerned, the first aspect that stands out is the aesthetic aspect of the landscape, through the differences in the dominant colour of the soil surface and the importance of the nature of the soil in the type of vegetation. The soil also preserves the memory and imprint of the passage of man and his activities over time. It also protects past bones, objects and buildings, and analyses enable archaeologists to reconstruct past climates and ecosystems, for example by dating charcoal or analysing organic molecules (W. CHRISTIAN et al. 2015). Finally, self-maintenance services encompass all the functions necessary for the production of other services: these include the water cycle, the nutrient cycle, primary production and soil formation (MEA, 2005) cited by W. CHRISTIAN et al. 2015).

The provision of all these ecosystem services may be at risk as a result of human activities and the climatic hazards that threaten them, leading to their deterioration. In fact, over the last 50 years, humans have modified ecosystems more rapidly and extensively than at any other comparable period in human history. These changes are largely linked to a rapidly growing demand for food, fresh water, timber, fibre and energy. The result has been a substantial loss of the Earth's biological diversity, much of it irreversible. These changes to ecosystems have led to the degradation of many ecosystem services (MEA, 2005).

According to the Millennium Ecosystem Assessment report, this degradation of ecosystem services could even increase significantly in the first half of this century, which would be a barrier to achieving the Millennium Development Goals.

What's more, as part of our quest to improve our well-being, we are developing activities that contribute to the further degradation of ecosystems and the services they provide. This can be seen in New Caledonia (Canada), where a project has carried out a study to assess the loss of ecosystem services linked to mining activity in the Grand Sud. Focusing on the method for converting the environmental impacts associated with mining activity into a loss of ecosystem services, based on mapping, a literature review and interviews, and the scenario-building method based on interviews, this study shows that the activity has negative impacts on New Caledonia's natural environments, leading to a deterioration in ecosystem services (RESCCUE PROJECT, 2016). These impacts include: stripping of vegetation cover; disruption of the hydrological regime; deterioration in water quality; emissions of CO2 and other greenhouse gases; and soil pollution by chemicals. These impacts can have repercussions on supply, regulation and even other services.

The effects of current changes in temperature and precipitation are also disrupting the various ecological functions in many ways. The main effects of climate change already observed are: extinction of species; modification of ecosystems; flooding of coastal areas; increase in cyclonic events; reduction in water resources, etc. These **are all having a major impact on the sustainability of current ecosystem services, thus affecting their production.** These all have major consequences for the sustainability of current ecosystem services, affecting their production. In extreme cases, they can lead to the collapse of ecosystems, sometimes within a very short timeframe (L. Bruno et al. 2015). Appropriate and urgent measures must therefore be taken to avoid this catastrophe.

1.1.2.2. Techniques and/or practices for managing and restoring land and ecosystem services

In order to curb the deterioration, studies have been carried out with a view to setting up a system, study and evaluate techniques for managing and conserving water and soil.

1.1.2.2.1. Techniques for managing land and ecosystem services

To address the negative impacts of land degradation and increase agricultural productivity in the long term, a wide range of land management practices have been developed over recent decades. According to R. WINTERBOTTOM et al (2013), four of these practices are the most promising for the drylands of sub-Saharan Africa. These are agroforestry, conservation agriculture, rainwater harvesting and integrated soil fertility management. Of these, agroforestry and rainwater harvesting are fast emerging as the key to increased agricultural productivity while restoring other valuable ecosystem services to agricultural landscapes. The authors explain these practices as follows:

1.1.2.2.1.1. Agroforestry

Agroforestry is a land-use practice involving the deliberate integration of perennial woody trees on the same plots as crops or livestock. Trees and shrubs in agroforestry systems can be regenerated and protected, or selectively planted and managed. Agroforestry systems can include both native and introduced non-native species. Where farmers include woody species that produce wood, fodder, edible leaves and other products, agroforestry systems evolve into more complex production systems that can provide a wider range of benefits and more resilient farming systems than those based on annual cereal production alone. The perennial woody trees used in agroforestry interact with soils and crops to create an agroecological system that enhances multiple ecosystem services and increases overall crop productivity.

1.1.2.2.1.2. Conservation agriculture

Conservation agriculture is a technique aimed at improving soil fertility and water use efficiency, especially on large farms. It is based on three related practices -Minimal soil disturbance through reduced tillage or the adoption of more environmentally friendly techniques. suitable for cultivation ;
-retention of crop residues or maintenance of a cover crop ;
-crop rotation or diversification of species grown in sequence and/or in association.
Combining these techniques stimulates the soil's organic matter content, retains soil moisture and protects arable land from erosion.

Adopting conservation agriculture increases rainfall infiltration, reduces runoff and pollution, improves soil fauna (micro-organisms) and sequesters soil carbon, as well as saving fuel and labour.

1.1.2.2.1.3. Rainwater harvesting

It covers a range of techniques used to trap rainwater before it runs off agricultural fields. These rainwater harvesting practices include: digging planting troughs or zaï; half-moons, which are crescent-shaped barriers dug into the earth; stone cordons placed around the edges; preparation of the soil in logs to form barriers or earthen trenches around the edges.

By slowing down run-off, these practices encourage infiltration, the development of plants, and so on. and also help farmers adjust to rainfall fluctuations.

1.1.2.2.1.4. Integrated soil fertility management

Integrated soil fertility management is another land and water management practice. to protect resources from degradation.
In its broadest sense, it refers to optimising the use of soil nutrient stocks, locally available resources and fertilisers to improve land productivity while maintaining or increasing soil fertility.

Its basic practices focus on the combined use of judicious but sufficient quantities of fertiliser and available soil conditioners. These soil improvers come mainly from local sources of organic matter (livestock or green manure, crop residues, compost or mulch, plant litter), but may also include lime and phosphate rock.These land management practices help to maintain agro-ecological balance, ecological resilience and the stability f ecosystem services, while ensuring subsistence and a variety of livelihoods for humans. If this management fails, leading to deterioration, restoration techniques can be adopted.

1.1.2.2.2. Techniques for restoring degraded land

According to E . ROOSEand G. DE NONI, (1998) reported by A. BOUBACAR NA-ALLAH, (2014), the Soil and Water Conservation (S.W.C.) is seen as the set of measures which, when developing natural resources, tend to maintain and/or increase production potential: soil and water being the fundamental elements of this potential. E. ROOSE, (1994) associates this notion of C.E.S with the question of soil fertility under the concept of (GCES): Gestion conservatoire des eaux et des sols, i.e. a new participatory strategy to respond to the problems of sustainable management of surface water, biomass and soil fertility. The effectiveness of these measures is justified by researchers who have carried out studies on this issue, which is of great concern to those involved in natural resource management.

In a study of erosion control methods in Algeria and Ecuador, E. ROOSE and G. DE NONI, (1998) have shown that the traditional techniques for managing surface water and land fertility developed by the ancients were certainly very effective in their era of stable climatic and socio-economic conditions, but are losing their effectiveness as human conditions change. Faced with this situation, modern equipment strategies have been adopted to curb this phenomenon. It is with this in mind that the water and soil management and conservation approach has emerged as a participatory rural development strategy. This strategy makes it possible to significantly increase land production and reduce the risk of erosion.

In the Sahel region of Burkina Faso, the increasing evolution of the soil degradation process is reflected in a reduction in plant cover or biomass, a decrease in the level of organic matter and in the biological activity of the soil, leading to physical and chemical soil degradation. Given this situation, J. BLAMAH (2004) conducted a study to assess the positive impact of anti-erosion techniques such as half-moons, subsoiling, filter dikes and stone barriers on the rehabilitation of degraded pastures. His studies show that the techniques studied create favourable conditions for the reappearance and maintenance of numerous forage species (herbaceous and woody), in that soil cover is improved by 3.5 to 40.5% depending on the type of management, and production was improved by 1.4 to 6.5 times compared with the control.

In Niger, meteorological conditions combined with human activities exacerbating the phenomenon of land degradation have prompted the country to experiment with a number of land protection and recovery techniques. It is in this context that O. LAMINOU MANZO et al (2020) conducted a study on the impact of benches in the recovery of degraded land in the rural commune of Simiri (Ouallam), which is subject to the most severe land degradation. The results of this study show that land reclamation works based on benches are effective because they have enabled a significant increase in the area covered by vegetation, and could even be an effective means of reducing climate risks through the development of ecosystem services provided by the development. Based on a study by I. BOUZOU MOUSSA and N. DAN LAMSO, (2004) in western Niger, they demonstrated that tassa (water pockets or basins) are also effective techniques for reducing runoff and erosion. They therefore help to avoid soil losses and ensure good agricultural production, especially when they are accompanied by other soil improvers such as manure and phosphorus, which further increases production.

Moreover, following the resurgence of this phenomenon in the Sahel in general and Niger in particular, projects have been set up to help increase production and restore natural resources, particularly soil and water. These include the project for integrated protection of agro-sylvo-pastoral resources in the department of Tillabéry Nord (PASP); the Tahoua rural development project (PDRT); and the rural development project for the Aguié district (PDRAA).

The aim of all these projects was to restore degraded natural resources by containing erosion and water loss through land reclamation and runoff collection works. In order to assess the effectiveness of these projects, G. YADJI and N. DAN LAMSO conducted a study in 2006 in the project areas. The results of their research showed that the structures built by these projects played an important role in restoring degraded resources. According to B. EDWIGE and R. CHRIS, (2009), these projects rehabilitated more than 250,000 ha of degraded land between 1985 and 2005.

In the Niger river valley at Niamey, the soil is subject to erosion, which strips thousands of hectares of the top layer of soil every year (M. KADER et al.2008). This exposes the soils to a loss of agronomic qualities, leading to a fall in yields. agricultural production, leaving many families vulnerable to food insecurity. In response to this threat of land degradation linked to erosion, techniques (mechanical and biological) for water and soil conservation and soil defence and restoration (CES/ DRS) were proposed and tested in partnership with local residents from surrounding villages. These practices have been a great success, as in 3 years they have enabled totally degraded soils to be recovered and put back into cultivation.

The reforestation system also plays an important role in restoring degraded land. In a study of the Sakey Koira Tegui site (Tillabéry), D. H. HALIDOU et al. (2020) showed that Acacia senegal is an appropriate species for reforesting the degraded soils on this site because of its resistance to the pedoclimatic conditions of this environment, which is characterised by a low organic matter content, a generally acid PH and a sandy texture.

1.1.3. Issues

For several years now, West Africa has been the focus of debate and research on environmental issues. For some time now, the Sahel-Saharan region has been characterised by a lack of rainfall, chronic droughts and rapid population growth, and is facing increasing degradation of its natural resources. In the Sahel, the periods of drought that have become more marked since the 1970s, combined with anthropic pressure resulting in land clearance, the extension of crop-growing areas, the reduction or disappearance of fallow land and overgrazing, have led to a major reduction in plant cover. These phenomena cause profound changes in surface conditions (S. ELHADJI MOUSSA, 2007). In Niger, this environmental concern is also currently emerging in the context of population growth and climate change, which have increased the area of exploited land and even the occupation of marginalised land (S. TAHIROU, 2017). This demographic growth is justified by the general population and housing censuses, according to which the population of Niger was 5,098,427 hbts in 1977;

7,500,000 hbts in 1988; 11,060,291 hbts in 2001 and 17,138,707 hbts in 2012. This same population is estimated to be 21,942,944 in 2019 (MINISTERE DU PLAN, 2020). Demographic growth, with its high demand for consumer goods and services and poor management, is putting great pressure on natural resources (soil, water, vegetation), leading to their deterioration. The climate, with its rainfall variations, violent winds and high temperatures, is also a threatening factor, as it results in water erosion through gullying and silting up of low-lying areas, and wind erosion through exposure of the soil to deflation. These processes lead to the widening of river banks, the displacement or disappearance of certain ponds, surface encrustation and above all irreversible degradation of the land. These erosion phenomena, or more precisely the degradation of the physical properties of the soil, are leading to a significant loss of land, which is becoming more and more widespread, especially in the western part of the country, which is facing heavy exploitation due to demographic pressure. In this part of the country, the strong anthropisation of the environment has led to land degradation phenomena characterised by the disappearance of plant cover, soil crusting, reduced fallow periods, etc. (S. ELHADJI MOUSSA, 2007; A. INGHATAN WARZAGAN, 2011).

The continuing degradation of natural resources in these semi-arid environments, resulting in a fall in agro-pastoral productivity, is making food security in these areas precarious. This situation is all the more worrying in that its effects are accentuated by the rapid growth of the population, causing an imbalance between ever-increasing needs and natural resources, which are steadily deteriorating (A. INGHATAN WARZAGAN, 2013). These soil degradation processes, which are characterised by a reduction in plant cover or biomass, leading to a reduction in the level of organic matter and biological activity in the soil, evolving in the form of a spiral, result in the physical and chemical degradation of the soil (J. BLAMAH, 2004). This phenomenon hampers the proper functioning of the ecosystem.

However, in view of this situation, which is getting worse by the day, some researchers have decided to take a closer look at it. have gone to great lengths to find solutions.
With this in mind, measures have been identified to curb this scourge.
Research in this area has identified a number of techniques that are more appropriate to the Sahel region, as they have already proved their worth in managing and restoring degraded land. In addition, to mitigate this degradation, nearly 17 billion CFA francs were invested in Niger to reclaim degraded land between 1980 and 1990, (B. YAMBA, 1995) reported by I. SEYNI MOUSSA, (2008). Remarkable technical, financial and human efforts are still being made to combat environmental degradation.

But despite the heavy investment of funds made to reclaim degraded land, this phenomenon persists throughout the western part of Niger in general, and the commune of Tondikandia in particular. This is a major socio-economic development issue. but also research. This is why this study focuses on the assessment of land degradation and ecosystem services in order to propose appropriate development measures.
To fully understand and study this phenomenon, the following questions are formulated around this issue.

1.1.3.1. Questions for research

1.1.3.1.1. The main question

What management methods can effectively combat land degradation and preserve ecosystem services in the Fandou Mayaki terroir?

1.1.3.1.2. Questions secondary

- What is the state of land degradation and ecosystem services in this area?
- What factors explain the state of land degradation in this area?
- What are the possible management alternatives for preserving land and ecosystem services?

1.1.3.2. The hypotheses of research

The following hypotheses are formulated for this study.

1.1.3.2.1. The main hypothesis

Sensible land management can reverse current trends and preserve land and ecosystem services in the long term.

1.1.3.2.2. Secondary hypotheses

- The degradation of land and ecosystem services is a major concern in this area.
- Land degradation factors may be due to climate variability, nature, the environment, or other factors. land use, farming methods and land management in the area.
- Possible management alternatives for preserving land and ecosystem services is through the adoption of good water and soil management and conservation practices.

1.1.3.3. The objectives

The objectives of this study are

1.1.3.3.1. The main objective

The main objective is to identify management methods that will effectively combat land degradation and consequently preserve ecosystem services.

1.1.3.3.2. Specific objectives

- First assess the state of land degradation and then that of ecosystem services in the study area.
- Characterise the factors that explain land degradation in this area.
- Propose alternatives and/or techniques for managing and preserving land and ecosystem services in the same locality.

1.2. PRACTICAL FRAMEWORK OR METHODOLOGY

This section summarises the different methods and tools used in this study in relation to the different research hypotheses.

1.2.1. Secondary hypothesis n°1

To assess land degradation in this terroir, the qualitative method was first applied, using a field observation technique. In fact, this field observation made it possible to learn about the current situation on the ground, particularly the degradation observed, by identifying the types of degradation and their extent.

This was carried out along two (2) east-west transects, one to the north and the other to the south of the reference village (Fandou Mayaki).

These transects were used to describe the different morpho-sedological units in the study area. These observations are illustrated by photographs, among other things, which reveal the reality of the terrain.

Then, in order to understand the threshold of this degradation, a synthetic map of the land degradation index was drawn up. This map is based on the method developed by P. BARBANT (2008). This consists of first presenting the types of degradation observed in the terroir, then estimating their extent at the scale of the land-use unit and then determining their degree. These last two indicators are combined to produce the synthetic land degradation index. All this was made possible by processing satellite images of the locality in 2020.

In addition, in order to assess the state of degradation of ecosystem services and to identify the current land management practices in this locality, the method used was a mixed method based on socio-economic surveys. These surveys focused on people's perceptions of the ecosystem services provided by natural resources, current land management methods, climate risks, their causes and their consequences for ecosystems. The surveys were carried out using a questionnaire (Appendix 1) sent to the heads of households in a simple random fashion. To this end, 62 questionnaires were administered and distributed as follows in the four surrounding villages (Fandou Mayaki, Kouboutché, Darey Beri and Gorma). A total of 62 people in the four villages in the study area were surveyed on the basis of a reasoned choice corresponding to 10% of households. In order to obtain relevant and reliable information in relation to the reference year, two criteria were adopted: The majority of the people surveyed should be aged between 50 and over. This criterion was chosen with the aim of surveying older people who have relevant retrospective information on the climate and that the respondent also has at least 30 years' experience on the study site, enabling him to observe the dynamics of his environment and the activities carried out there in the context of climate variability and demographic growth.

This made it possible to gather qualitative and quantitative information, which is processed using Sphinx and Excel software. This information is sensitive because it is the testimony of the people concerned themselves. Table 1 shows the situation in the villages surveyed, and Figure 1 shows the age range. of the people surveyed.

Table 1: Breakdown of households surveyed in the study villages

Villages surveyed	Number of farm households in 2012	Number of households surveyed per village	Percentage of sample by village
Fandou Mayaki	345	35	56,45
Kouboutché	136	14	22,58
Darey Beri	72	7	11,29
Gorma	65	6	9,68
Total	618	62	100

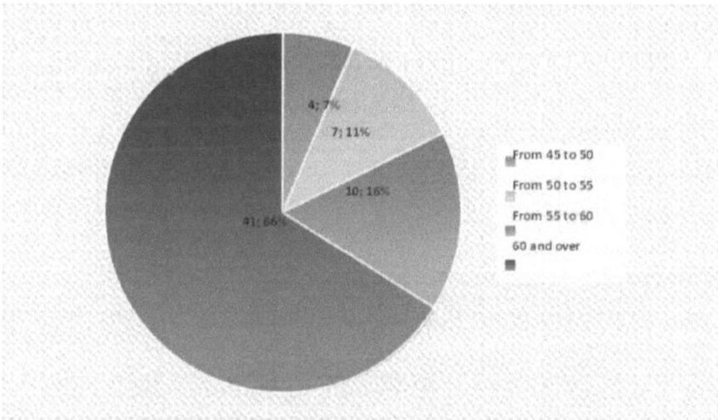

Figure 1: Age range of people surveyed

Figure 1 shows that 93% (58/62) of the people surveyed were aged between 50 and over and had been living on the study site for 30 years or more. This made it possible to gather sensitive information from older people with a length of experience that enabled them to observe the dynamics of their environment and their activities in the context of climate variability and population growth.

The tools used are :

- an observation sheet (Appendix 2) describing the different morpho-pedological units, the types of damage observed in the field and the different plant formations;
- Arcgis10.1 and ecognition developer 64 software to produce the various land degradation maps;

- a mobile phone used to photograph the various observation points and take contact details.

- an individual survey form addressed to the heads of agricultural households.

1.2.2. Secondary hypothesis 2

To characterise the land degradation factors in the area, rainfall and grain size data, surveys and satellite images were processed. The first step was to determine the average annual cumulative rainfall recorded from 1988 to 2018 at the Filingué weather station and the standard deviation, and then to calculate the overall average rainfall for the series in order to arrive at the standardised rainfall index. This index, which reflects a rainfall surplus or deficit for the year in question compared with the reference period, is calculated on the basis of the formula of P. LAMB, (1982) which is equal to: (Xi - X)/S with Xi: rainfall in year i, X: average interannual rainfall over the reference period and S: standard deviation of interannual rainfall over the reference period.

In addition, the number of rainy days per year at the same station is determined and analysed to generate figures correlating average annual rainfall and average annual rainy days. This makes it possible to explain the variations observed between different years. The data were collected at the Filingué departmental agricultural office and processed using Microsoft Office software (Excel).

Soil samples were then taken for granulometric analysis at the geomorphology laboratory in order to highlight the characteristics of the soils in the study area. A total of ten (10) samples were taken from 20 to 30 cm below the surface, from observation transects. To help identify the samples, the northern transect is labelled T1 and the southern transect T2.

The choice of these transects results not only from the layout of the villages in the different geomorphological units, but also from the geomorphology of the terrain itself.

For the analysis, a 100 g mass of each sample was taken and then sieved wet on a 0.04 mm square-mesh sieve. This wet sieving was used to separate the coarse fraction of grains with diameters greater than 0.04 mm from the fine fraction with diameters less than 0.04 mm. The coarse fraction retained was dried in an oven at 105°, then weighed on a precision balance to the nearest thousandth of a gram.

The rejected material, dried and weighed in this way, was dry-sieved on a column consisting of seventeen (17) square-meshed sieves with the following edges: 2 mm; 1.6 mm; 1.25 mm; 1 mm; 0.8 mm; 0.63 mm; 0.5 mm; 0.4 mm; 0.315 mm; 0.25 mm; 0.2 mm; 0.16 mm; 0.125 mm; 0.1 mm; 0.08 mm; 0.063
mm; 0.05 mm and 0.04 mm. The reject from each sieve was then weighed and the mass recorded for statistical purposes. These mass data from the granulometric analyses were used to produce relative and cumulative frequency curves. This made it possible to characterise the sandy texture of the samples from each transect. A range of particle size analysis equipment (sieve, beaker, oven, balance) was used to carry out this analysis.

Maps of land use were produced by processing satellite images. These were first downloaded from the Earthexplorer site, then cut according to the study area and exported to Ecognition developer 64 software for segmentation and classification of the different land use units. It was possible to identify the different units using Niger's 2001 land-use nomenclature and knowledge of the terrain. After this phase, the images were re-exported to Arcgis 10.1 for correction and meshing. To identify land use and management methods, data was collected through surveys.

1.2.3. Secondary hypothesis n°3

For the last hypothesis, which would consist of proposing possible management alternatives for preserving land and ecosystem services, documentary research is of capital importance in order to pertinently analyse and clearly explain the effectiveness of the various techniques for managing, restoring and conserving degraded land. It was therefore carried out in documentation centres and services such as the geography department library, the FLSH library, the Faculty of Agronomy library, the AGRHYMET library and the CNEDD as well as on the Internet and also from resource persons who can provide information or data relating to this issue. The common equipment used for these different methods is the laptop, which is used to all activities.

1.3. PRESENTATION OF THE STUDY AREA

This section presents the geographical location and the various biophysical and socio-economic characteristics of the study area.

1.3.1. Location

The area covered by this study is located in the rural commune of Tondikandia, which is one of the communes making up the department of Filingué (Tillabéri region). It covers an area of approximately 2850 km2 and is essentially made up of sandstone plateaux of the terminal Continental, cut by dry fossil valleys carved out at the beginning of the Quaternary (B. YAMBA, 1989).

This Commune is located between 2° 30' and 3° 50' longitude East and 13° 40' and 14° 20' latitude North. It is bordered to the north by the communes of Kourfey Centre, Imanan and Filingué; to the south by the communes of Simiri and Dingazi Banda (Ouallam); to the east by the department of Loga and to the west by the commune of Tagazar (Balleyara). This commune is divided into six (6) sub-zones, including the Attaloga zone, the Damana zone, the Fandou Mayaki zone, the Kandoum zone, the Talifanta zone and the Tibewa zone (PDC Tondikandia, 2020). The Fandou Mayaki zone is the subject of this study and is made up of 28 administrative villages, 4 of which were surveyed. These villages lie between meridians 2°53' and 2°56' and parallels 13°52' and 13°56'.

Map 1 shows the commune of Tondikandia and the villages surveyed.

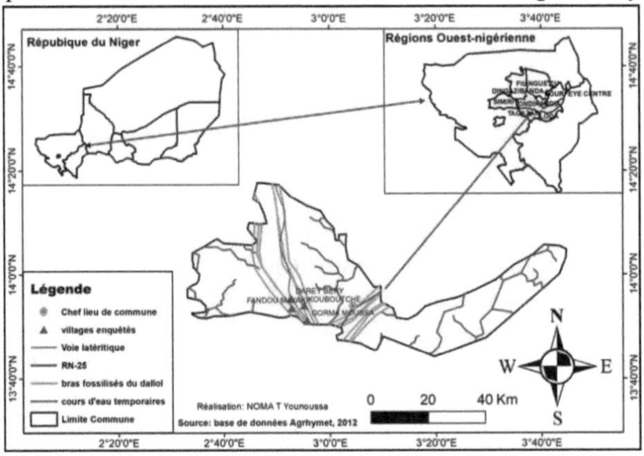

Map 1: Location of the study area

1.3.2. Biophysical characteristics of the environment

1.3.2.1. Geology

Geologically, the Tondikandia commune is located in western Niger, in the great sedimentary basin known as the Iullemeden Basin. This sedimentary basin is a vast syncline (dipping 1 to 2%) covering the western part of the Republic of Niger. It is open to the north-west by the Taoudenni basin (Mali), bounded to the north by the Aïr mountains, to the south and south-west by the Liptako basement, to the east by the Damagaram Mounio and to the south by the Nigerian basement. The geological history of this basin is marked by marine transgressions interspersed with continental episodes from the Palaeozoic to the Cenozoic (GREIGERT, 1966) quoted by M. MALAM ABDOU, (2005).

In fact, it is essentially made up of Tertiary terminal continental formations, which are detritic sediments dominated by clay and sandstone that have filled the basin to a depth of more than 450 m (DUBOIS and LANG, 1984) reported by M. MALAM ABDOU, (2005). These sedimentary formations are cut by dry fossil valleys dug at the beginning of the Quaternary. These valleys, whose presence breaks up the monotony of the landscape, are filled in by deposits. alluvial and aeolian zones and occupied by temporary watercourses (B. YAMBA, 1989).

1.3.2.2. The landscape units

From a geomorphological point of view, the Iullemeden sedimentary basin is characterised by very extensive sandstone plateaux with armoured undersides. These plateaux, with an average altitude of 260 m, have steep slopes that are dissected by fossil hydrographic networks tributary to the River Niger (the Dallols) (M. MALAM ABDOU, 2005).Thus, according to F. OUSSEINI SIDIKOU, (2015) observation and analysis of the topographic map of the area (cf.: topographic map, sheet ND-31-VXI), makes it possible to distinguish four types of geomorphological units in the commune of Tondikandia of varying proportions.

1.3.2.2.1. The setbacks

The plateaux cover a larger part of the commune, i.e. 33% of its total surface area (F. OUSSEINI SIDIKOU, 2015). The underside of the plateaux is made up of a light eolian scattering, covering laterite or ferruginous cuirasse, which gives the soils a gravelly appearance (O. AMADOU, 2016) taken from SOS Sahel, 2002. On these plateaux, the soils also belong to the group of tropical ferruginous soils with little or no leaching, with a sandy to sandy-clay texture, generally poor with a very low humus content rarely exceeding 2% (B. YAMBA, 1989) and F. OUSSEINI SIDIKOU, 2015). These plateaux are generally flat or slightly undulating. On the tops of these plateaux there are small depressions of around ten square metres where run-off water can stagnate and form a sort of permanent pond.

The following facies can be distinguished on these plateaux: tiger bush, cultivated areas with sparse vegetation and deeply denuded areas (Y. MOUKAILA, 1988).

From an agricultural point of view, the plateau tops do not offer much potential. However, they are used as grazing areas and provide most of the firewood and timber for farmers. As a result, these geomorphological units are now completely bare. These plateaux are bounded by rocky or scree slopes, or by outwash glacis that have been heavily gullied in their upstream parts (O. AMADOU, 2016).

1.3.2.2.2. The slopes of plateaux

The slopes of the lateritic plateaux cover 19% of the area of the commune (F. OUSSEINI SIDIKOU, 2015).They occupy areas where erosion by washing and gullying on the steepest slopes is the most important pedogenetic phenomenon. These slopes can be subdivided into upper slopes where the initial geological material outcrops and lower slopes where this geological material is covered by eolian or colluvial deposits. Many boulders, pebbles and lateritic gravels are currently outcropping on these units. This situation means that these areas are marginalised and reserved solely for livestock grazing (B. YAMBA, 1989; F. OUSSEINI SIDIKOU, 2015 and O. AMADOU, 2016).

1.3.2.2.3. Glazing

The glacis, made up of aeolian systems (dune, dune complex) occupying the western part of the municipality and alluvial systems (valley, dallol and shallows) reworked on fairly flat surfaces with gentle slopes of around 1 to 2%, cover 47% of the municipality (F. OUSSEINI SIDIKOU, 2015; O. AMADOU, 2016). Most of these glacis have been degraded upstream by gullying and deflation, and have variable contours: gently undulating surfaces between gullies sometimes bare and studded with thin sandy-loam films (runoff and desiccation crust), sometimes silted up with fine, mobile aeolian material or <<bad-lands>> with claws and deep gullies in which coarse elements from the slopes and remobilised by torrential flows accumulate, whereas the downstream area is sandy and generally functional. These glacis are largely made up of ferruginous soils with little leaching, essentially sandy (O. AMADOU, 2016).

1.3.2.2.4. The shallows

The lowlands that make up the depression system occupy 1% of the commune (F. OUSSEINI SIDIKOU, 2015).

These are the lowest landscape units in the toposequence and contain the areas where water from the slopes meets and organises itself (O. AMADOU, 2016). These most depressed areas are often temporarily flooded, where rainwater causes silting by spreading. In addition, the valley bottoms constitute intermittent drainage channels, occupied in the clay basins by semi-permanent pools extending in the direction of the valley.

They are generally wide and elongated and made up of clayey and sandy-loam materials with remarkable plant cover (B. YAMBA, 1989; O. AMADOU, 2016).

These ponds are surrounded by dense vegetation of species that tolerate temporary hydromorphy, such as Mytragina inermis, Accacia nilotica and Piliostigma reticulatum, and offer significant potential for market gardening, especially in the Damana zone (B. YAMBA, 1989).

1.3.2.3. Hydrography

The Iullemeden basin is characterised by a hydrographic network known as the dallols that drain the region. These dallols, which were once functional, are now completely degraded and very often become centripetal and endoretic, disconnected from each other. In fact, this situation could be the result of obstructions caused by localized solid deposits, particularly at the junction of the drain and the collector, or by the erosion of the bed slope (by aeolian and colluvial sand deposits) of the main collectors, thus giving rise to several hydrographic sub-networks (DESCONNETS, 1994 reported by M. MALAM ABDOU, 2005). This gives rise to a discontinuity in flows, resulting in a string of temporary or permanent pools during the winter months.

The commune of Tondikandia, which is the subject of this study, also has significant surface water resources. Thus, according to the diagnosis of the Commune's natural resources, in 2014 there were forty (40) permanent ponds and sixty-two semi-permanent ponds spread across the central part of the Commune (F. OUSSEINI SIDIKOU, 2015). In addition, a seasonal watercourse (branch of the dallol bosso) crosses the commune during the rainy season from the Malian border to the Balleyara confluence, passing through Wankama to feed the River Niger in its middle section. The Fandou area lies in the valley of one of the branches of the dallol bosso, now fossilised by alluvial and colluvial deposits.

1.3.2.4. Climate

In Niger, as in the rest of West Africa, atmospheric circulation is controlled by the position and relative importance of anticyclones and thermal depressions. This atmospheric circulation is marked by the alternation of two air masses (the harmattan and the monsoon) of different origin and character. The first is hot, dry continental tropical air. This is an air mass that blows in from the north or north-east, originating from thermal depressions in the Sahara. The second is a relatively cold and humid air mass originating from the anticyclone. of Sainte-Hélène in the Atlantic Ocean (Y. MOUKAILA, 1988; M. MALAM ABDOU, 2005).

The contact surface between these two masses forms what is known as an intertropical front, whose movement along the meridian determines two seasons in Niger: a dry season and a winter season (M. MALAM ABDOU, 2005).

The climatic conditions in the department of Filingué, as in Niger in general, are characterised by a tropical Sahelian climate marked by a long dry season running from November to June and a relatively short rainy season from June to October. Rainfall in this area is irregular and unevenly distributed in time and space, with annual averages of around 650 mm/year between 1931 and 1970. Since 1968, however, there has been a decline in rainfall, with more frequent periods of drought such as 1968, 1971, 1974, 1982, 1984 and 1987 (Y. MOUKAILA, 1988). This is confirmed by F. OUSSEINI SIDIKOU, (2015) according to whom the average annual rainfall at the Filingué station was around 400 mm/year between 1950 and 2013. The climate in this area is also characterised by a hot, dry wind (harmattan) that blows from November to April in a northerly or north-easterly direction towards the south-west, and a hot, wet wind (monsoon) that generally blows from May to October in a southerly or south-westerly direction towards the north-east (F. OUSSEINI SIDIKOU, 2015).

On the basis of all these characteristics, we can assume that the commune of Tondikandia, included in this department, some sixty kilometres from its capital (Filingué) shares the same climatic characteristics as the latter, as shown in Figure 2.

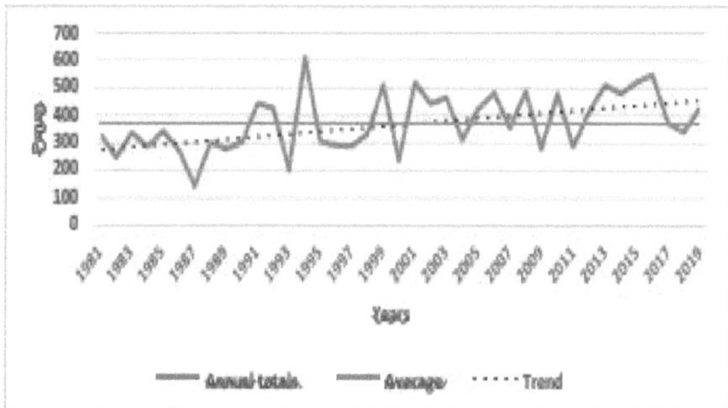

Figure 2: Interannual change in rainfall in the Tondikandia commune from 1981 to 2019, Source: **Tondikandia PDC, 2020**

This graph shows that rainfall varies spectacularly in this municipality, with a large gap between the maximum of 607.8 mm in 1994 and the minimum of 135.1 mm in 1987, while the inter-annual average is 370.16 mm.The figure shows three (3) different periods: from 1981 to 1990, when rainfall fell significantly; from 1991 to 2000, when rainfall varied between wet and dry years; and from 2001 to 2019, which is the wettest period in the series, although it also contains some dry years. It should be noted that this graph shows inter-annual variability in rainfall, with alternating wet and dry years between 1991 and 2000, in contrast to the persistence of dry years before 1990 and wet years between 2000 and 2019.

1.3.3. Demographic and socio-economic characteristics

1.3.3.1. Population trends and characteristics

The rural commune of Tondikandia, in which the study area is located, is experiencing the same demographic trends as most other rural communes in Niger. According to INS statistics, the population of this commune was 108147 in 2012. With a regional annual growth rate of 3.2% (INS, 2015), this same population is estimated at 145171 inhabitants in 2020 and will rise to 168255 inhabitants by 2024 (INS, 2021).

With a surface area of around 2,850 km², it is characterised by strong demographic growth, with high densities ranging from 37.94 inhabitants per km² in 2012 to 50.93 inhabitants per km² in 2020, rising to 59.03 inhabitants per km² in 2024 (Table 2). This demographic explosion is putting great pressure on natural resources (land, water, vegetation, etc.), which are dwindling by the day. This has socio-environmental consequences, including a shortage of arable land, which is already a reality in most of the villages in the commune, a reduction in the length of fallow periods, a deterioration in the plant cover, a fall in agricultural production, an increase in the rural exodus to large towns such as Niamey or to neighbouring countries, thus emptying the villages of their able-bodied people, etc.

Table 2: Changes in population size and density in Tondikandia from 2012 to 2024; Source: INS, 2021

Years	Population	Density
2012	108147	37,94
2013	112206	39,37
2014	116414	40,84
2015	120777	42,37
2016	125301	43,96
2017	129994	45,61
2018	134865	47,32
2019	139922	49,09
2020	145171	50,93
2021	150621	52,84
2022	156275	54,83
2023	162150	56,89
2024	168255	59,03

1.3.3.2. Socio-economic activities

The commune's main socio-economic activities are farming and livestock rearing, which are practised by over 80% of the population. Other activities include handicrafts, trade and migration, which are considered to be secondary activities in rural areas.

1.3.3.2.1. Agriculture

Agriculture remains the main activity practised by almost all of the commune's inhabitants. In fact, over 90% of its sedentary inhabitants derive their livelihood from agriculture, in other words, it is the main staple food and source of income for the commune's population.
Arable land is sown with two types of crops: rain-fed and irrigated.

1.3.3.2.1.1. Rain-fed agriculture

It is the main form of farming practised in all the villages of the Tondikandia commune. It occupies the majority of the commune's population. This activity lasts throughout the rainy season. The main rainfed crops are millet, sorghum, beans, groundnuts, sorrel, sesame, okra, voandzou, souchet, etc., which are food and cash crops.
Millet is grown in the dune fields and on the slopes of certain plateaux; it is often combined with beans. Sorghum is grown in the valleys. Okra, groundnuts and other crops are grown on plots of land in the fields by women.
This type of farming is characterised by :

- heavy dependence on irregular rainfall that is often poorly distributed in space and time
- the disappearance of fallow land as a method of restoring fertility to fields, given the shortage of agricultural land;
- Overexploitation of farmland, which is increasingly leading to soil degradation, resulting in lower production;
- the use of rudimentary techniques that do not ensure satisfactory production;
- farming without the use of appropriate inputs.

As a result, almost every year there are major cereal deficits, leading to repercussions on the nutritional status of populations (PDC Tondikandia, 2020).

1.3.3.2.1.2. Irrigated agriculture

Irrigated or off-season farming is practised in the commune on the dallol and lowlands. The main crops are legumes, fruit and tubers. This makes it both a consumer and a commercial activity, with significant nutritional benefits for the local population. This activity represents an important source of income for those who practise it. It also helps to reduce the cereals deficit in rain-fed agriculture. With the cereal deficit of recent years, the existence of semi-permanent ponds and the installation of standpipes everywhere, the practice of this activity is expanding in this municipality.However, irrigation methods are still archaic, with little use of motor pumps, improved seed and other modern materials. This hampers the contribution of this activity to the development of the commune.

1.3.3.2.2. Breeding

Livestock farming is the commune's second main activity after agriculture. It is practised by both men and women, and accounts for over 80% of the working population. Livestock farming and agriculture complement each other. In fact, almost every household has livestock. These animals include azines, cattle, camels, goats, horses and sheep, the most popular of which are goats, cattle and sheep.
There are two types of livestock farming in this locality: sedentary and transhumant. Sedentary livestock farming mainly concerns small ruminants, which are reared at home by agropastoralists to meet their primary needs. This type of farming provides savings in the event of a financial emergency for food, ceremonies or other activities. Cows are kept for their milk and donkeys and oxen for transport. As for transhumance, this is made up of the rest of the herds, especially the large ruminants, entrusted to shepherds who take them to pasture far from the villages towards the pastoral zone (North). However, this type of

livestock farming is in decline because of the insecurity that has taken hold in the country, especially in the northern part of Tillabéry, which is a pastoral area.
These animals graze on the plateaux and uncultivated glacis land.

Animal feed is essentially based on rangeland resources, aerial fodder and crop residues. Animals are watered in ponds, water reservoirs and modern watering points (pastoral wells and pumping stations) (PDC Tondikandia, 2020).
The population is also involved in poultry farming. The main species are chickens and guinea fowl, to which are added ducks and pigeons. The products and by-products of poultry farming are used for food and, above all, for sale to satisfy other needs of the local population. households (food, clothing and certain religious ceremonies). They also provide organic manure to fertilise the soil.

1.3.3.2.3. Craftsmanship

Crafts are practised in the commune, but to a limited extent. There are a variety of craft trades, including blacksmiths, tanners, tailors, shoemakers, bricklayers, potters, welders, mat makers, carpenters, mechanics, hairdressers, braiders, vulcanisers, well makers and butchers. Most of these craftsmen work on their own, in an archaic and periodic fashion. However, this activity, which could contribute to satisfying primary needs, is hampered by a lack of technical and financial support, a lack of training for craftspeople, difficulties in selling craft products and access to raw materials.

1.3.3.2.4. The trade

Trade generally concerns agro-sylvo-pastoral products, and transactions are much more significant after the harvest. Trade mainly involves staple products such as millet, sorghum, beans, lettuce, cabbage, condiments, etc., which are bought and sold at local markets.
These are followed by handicrafts and manufactured products such as mats, clothes, shoes, sugar, tea, foodstuffs, etc. All these exchanges take place at the large weekly markets in the commune, which are the Damana, Fandou Mayaki, Tibewa, Fandou Beri and Attaloga markets. They are based on local, imported and exported products. Despite all this, this activity is not well developed in the commune. This is mainly due to the isolation of many villages and the advanced deterioration of tracks in certain localities. This makes it difficult to sell local agro-sylvo-pastoral products and to supply markets with other products.

1.3.3.2.5. The exodus

In the current context, the exodus is seen as a practice aimed at strengthening the resilience of households in the face of climatic hazards and the scarcity of arable land. In fact, this is a very common practice in this commune, with almost all the population leaving their village just after harvest time for neighbouring towns. The reasons for this exodus are quite simply economic, as those who leave not only reduce their family burdens, but above all support their families by sending money to make up the food shortfall.

1.4. DIFFICULTIES ENCOUNTERED

Any work undertaken is not without its inherent vices, especially for a beginner. A number of difficulties were encountered during the course of this study. The fact that the data collection period coincided with the rainy season meant that meeting the people surveyed was quite a problem. In addition, the lack of security and the ban on motorbike traffic prevented us from accessing certain remote areas for the observation phase, which is why the entire investigation was only carried out in the eastern part of the zone. The fact that we were not fully conversant with GIS software also led to anomalies in the production of the various maps. Finally, the lack of access to certain data on other climatic parameters led to a failure in the simulation of variations in climatic parameters.

CHAPTER 2

STATE OF LAND DEGRADATION AND ECOSYSTEM SERVICES

This chapter will first assess the state of land degradation and then the dynamics of ecosystem services in the study area.

2.1. ASSESSMENT OF THE STATE OF LAND DEGRADATION IN THE LAND OF FANDOU MAYAKI

Land degradation is a very significant phenomenon in the Fandou Mayaki terroir. It takes several forms, including erosion and structural degradation. The most salient types in this locality are water and wind erosion, and physical and biological degradation. Chemical degradation could not be determined due to a lack of the necessary equipment. However, it can also be seen in the drop in agricultural production, as the respondents stated. The extent of these types of degradation varies from one unit to another. It ranges from slight to very severe. As for the degree, it ranges from low to very high. The combination of these two factors made it possible to determine the degradation index, which varies from low to very high. Photos 1, 2, 3, 4 and 5 show the situation observed in terms of degradation in this locality. Maps have also been drawn up to confirm this hypothesis.

Photo 1: A completely encrusted area at the foot of the Kouboutché plateau

Photo 2: A large rock outcrop on the Kouboutché plateau

Photo 3: A surface stripped bare by deflation on the downstream glacis (T1)

Photo 4: An area under severe threat from gully erosion at the foot of the Gorma plateau

Photo n°5 : Silting in the Gorma shallows

There are several methods for assessing land degradation. For this study, the method of P. BARBANT, (2008) is adopted. This method consists of identifying the various degradation indicators, using which an index will be determined to characterise the overall state of land degradation. Cartography is essential for this, in order to produce maps of types, extent and degree of degradation. The combination of these last two indicators has led to the production of a synthetic land degradation index map.

2.1.1. Types of land degradation

The different types of degradation observed in the Fandou Mayaki terroir are wind erosion, water erosion, biological degradation and physical degradation. These types of degradation vary according to the land-use units. For degraded scrub, degraded tiger scrub and riparian forest units, the most dominant types are wind erosion (deflation), water erosion (sheet, linear and gully erosion) and biological degradation (loss of vegetation). Regular riparian formations are also subject to the same types of degradation, with the exception of biological degradation. Human dwellings are characterised by water erosion (gullying in neighbourhood streets). Fallow land and cultivated areas are threatened by biological degradation (loss of vegetation), physical degradation (crusting), wind erosion (deflation and formation of nebkas) and water erosion (diffuse erosion). In bodies of water, wind and water erosion lead to silting. Bare surface units suffer the effects of physical degradation through crusting, water erosion through diffuse and linear erosion, wind erosion through deflation and biological degradation with a huge loss of vegetation. The rocky terrain is marked by wind erosion (deflation), water erosion (diffuse erosion) and biological degradation (loss of vegetation). Map 2 shows the current state of the types of degradation observed in the Fandou Mayaki terroir.

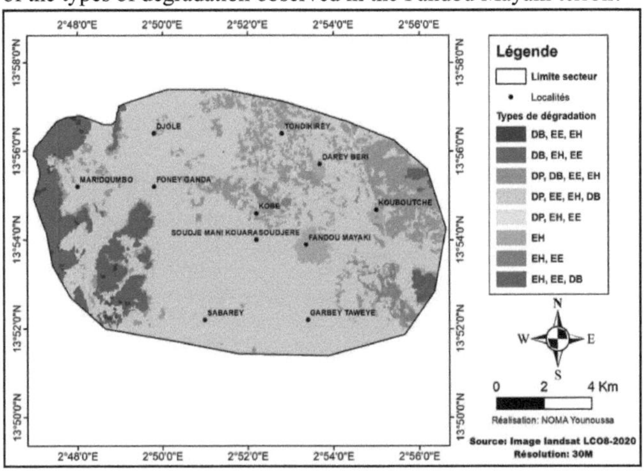

Legend: EE: wind erosion; **DP**: physical degradation; **EH**: water erosion; **DB**: biological degradation

Map 2: Types of degradation observed in the Fandou Mayaki terroir.

2.1.2. The spread of land degradation

For the purposes of this study, the extent of degradation refers to the surface area occupied by the type of degradation at the scale of the land-use unit. By estimating the extent of degradation in this terroir, we were able to determine four classes of extension. The low, medium, high and very high extension classes. These different levels of extension of degradation vary from one land use unit to another. Thus, the low class, which corresponds to between 5% and 25% of the degraded surface area of the unit, is observed in human settlement units, fallow land and some cultivated areas in the village halo. The medium extension class corresponds to degradation occupying between 26% and 50% of the unit. This extension class is found on units with regular riparian formations, bodies of water and farmland. For the strong extension class, it indicates that degradation occupies a surface area of 51% to 75% of the unit.The latter can be seen in the degraded scrub, degraded tiger scrub, degraded riparian formations, rocky terrain and cultivated areas units. As for the extension class very high, representing areas where more than 75% of the surface area has been degraded. This is observed on the bare surface units, which as a whole are encrusted.

Map 3 shows the extent of degradation according to different units occupancy of land use.

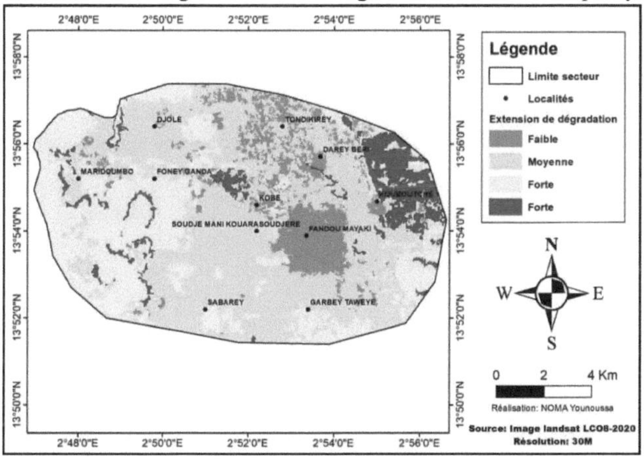

Map 3: Extent of land degradation in the Fandou Mayaki terroir.

2.1.3. The degree of land degradation

According to P. BARBANT (2008), the degree of degradation is the stage of seriousness (or severity) reached by a given type of degradation in a given area of land. It is assessed using two methods: one that refers to physical soil variables and another that mainly refers to the impact of degradation on land productivity. The application of these methods made it possible to determine four classes of degree of degradation in the Fandou Mayaki terroir. These are low, medium, high and very high. Table 3 shows the degree classes and the surface area occupied by each in the study area.

Table 3: Surface area of land degradation classes in the Fandou Mayaki terroir, Source: our study

Degree of degradation	Surface area (ha)	Proportion (%)
Low (2)	1779,48	11,56
Medium (3)	7252,47	47,09
Strong (4)	3582	23,26
Very strong (5)	2785,95	18,09
Total	15399,9	100

This table shows that the medium grade occupies the largest area of land, i.e. 47.09%. This is followed by the high-grade class, which accounts for 23.26%. This is followed by the very strong class, which accounts for 18.09% of the study area. The weak class comes last, representing just 11.56% of the area. These degree classes, like the previous indicators, differ from one occupation unit to another. In fact, the low degree of degradation characterises fallow land, human dwellings and cultivated areas close to dwellings. Units with a medium degree of degradation are other cultivated areas that are heavily exploited but some of which are altered by the farmers. The high degree of degradation is observed in degraded, regular riparian formations and bodies of water. For degraded scrub, degraded tiger scrub, bare surfaces and rocky terrain, the degree of land degradation is estimated to be very high. This is shown on map 4. It shows the classification of this degree on the various units.

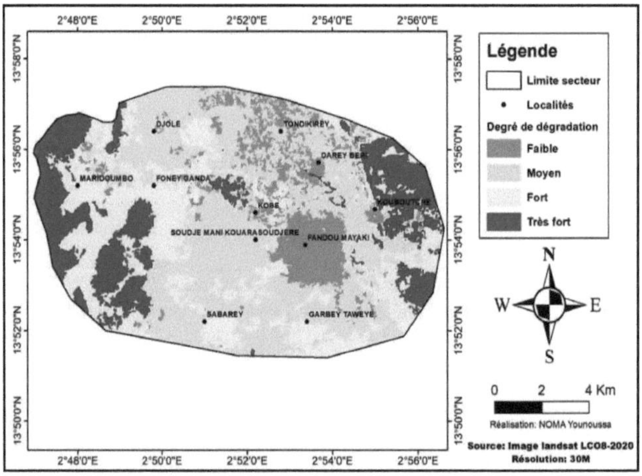

Map 4: Degree of land degradation in the Fandou Mayaki area

2.1.4. The synthetic land degradation index

The synthetic degradation index is the main parameter for assessing land degradation. It is determined on the basis of two indicators: the extent and the degree of the types of degradation. This parameter makes it possible to characterise the threshold of land degradation from one environment to another. Table 4 shows the different classes of index determined in the Fandou Mayaki terroir.

Table 4: Areas covered by land degradation index classes in the Fandou Mayaki terroir, Source: our study

Degradation index	Surface area (ha)	Proportion (%)
Low (2)	1779,48	11,56
Medium (3)	7124,22	46,26
Strong (4)	3710,25	24,09
Very strong (5)	2785,95	18,09
Total	15399,9	100

This table shows four (4) indices for this terroir: the low index, the medium index, the high index and the very high index. It can be seen that moderately degraded land in this locality represent almost half of the study area, with a surface area estimated at 7124.22 ha, i.e. 46.26% of the land. However, they are characterised by severe chemical degradation, as shown in Table 6, which shows that farmers have seen a sharp drop in agricultural production in this area over the last 20 years. This is followed by high index land, i.e. land that is severely degraded. These cover a surface area of 3,710.25 ha, with a proportion of 24.09%. Very severely degraded land comes third, covering 2,785.95 hectares, or 18.09% of the area. Weakly degraded land covers a large part of the sector, with an area of 1,779.48 ha, or 11.56%. These land degradation classes are identified on the different land use units. As shown on Map 5, human habitation units, fallow land and some cultivated areas have suffered little degradation. Moderate degradation is evident over almost the entire area. The cultivated areas, although heavily exploited, are less degraded than the other units. The degraded riparian formations, the regular riparian formations and certain cultivated areas make up the units where degradation is considered to be high. The units that have suffered very severe degradation of their land are degraded scrub, degraded tiger scrub, bare surfaces and rocky terrain. Map 5 shows the current state of land degradation in the Fandou Mayaki area.

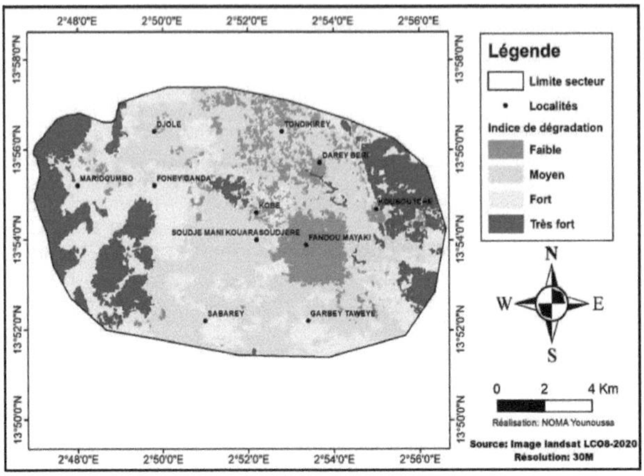

Map 5: Synthetic land degradation index in the Fandou Mayaki area

2.1.5. Summary of the state of land degradation in the Fandou Mayaki area

The overall state of land degradation in this area is shown in Table 5. This table is made up of the various indicators and the synthetic land degradation index. The table shows that each land use unit is subject to at least two types of degradation, with the exception of human dwellings. This shows that all the units are subject to several types of degradation. In addition, this table presents the extension classes of land degradation types on the different units. Thus, two (2) units have an extension of types of degradation judged to be low and two (2) others whose extension of degradation is estimated to be medium. Three (3) units have degradation extension estimated to be high to very high, while all three classes (low, medium and high) are observed on cropland. Apart from the extent, the degree of degradation is also presented in this table. The degree of degradation varies from low to very high, with the degree of degradation of human dwellings and fallow land considered to be low. The average degree is estimated for a single unit, while the high degree is observed in only four units. The very high level is also assessed on four land use units. Finally, this table shows the synthetic index resulting from the combination of these two parameters. This is the and this index shows that degradation is estimated to be low on two units. It also shows that this same deterioration is considered weak, medium and strong on one unit, while it is rated as strong to very strong on seven remaining units.

Table 5: Summary of the state of land degradation in the Fandou Mayaki terroir, Source: our study

Land use units	Dominant types of degradation	Class extension	Class of degree of degradation	Total value	Value of the index	Qualification index
Degraded scrub	ErosionWind erosion, erosion water erosion, biological degradation	4	5	9	5	Very strong
Degraded tiger bushes	ErosionWind erosion, erosion water erosion, biological degradation	4	5	9	5	Very strong
Degraded riparian formations	Physical degradation, erosion erosion, erosion wind erosion, biological degradation	4	4	8	4	Fort
Regular riparian formations	Physical degradation, water erosion erosion, wind erosion	3	4	7	4	Fort
Human settlements	Water erosion	2	2	4	2	Low
Fallow land	Biological degradation, physical degradation, wind wind erosion, water erosion	2	2	4	2	Low
Bodies of water	Erosionwind erosion, water erosion	3	4	7	4	Fort
Bare surfaces	Physical degradation, erosion erosion, erosion wind erosion, biological degradation	5	5	10	5	Very strong
Rocky ground	Erosionerosion, erosion wind erosion, biological	4	5	9	5	Very strong

	degradation					
Zonescul tivation	Physical degradation, wind erosionerosion, erosion water erosion, biological degradation	2 3 4	2 3 3	4 6 7	2 3 4	Low Medium Fort

2.2. EVALUATION OF THE DYNAMICS OF ECOSYSTEM SERVICES IN THE FANDOU MAYAKI AREA

2.2.1. Procurement services

2.2.1.1. Food resources

2.2.1.1.1. Agriculture

Agriculture is the most important of these services. There are two types of farming: rain-fed and irrigated. Rain-fed agriculture is the most common form of farming in this locality, as shown in Figure 3. Of the 62 people surveyed, 100% said they farmed during all the reference periods. However, in recent years, climatic variations combined with other factors such as land degradation have meant that this activity has not been able to keep up with demand. to meet the needs of the population. As a result, it has to find other solutions to make up for the shortfall in agricultural production, hence the intensification of irrigated farming or off-season cultivation. This activity was less popular among the population 30 years ago, as only three (3) people grew lettuce, cabbage and potatoes as their main crops. Ten years on, we can see that this activity has evolved, with more than five million people taking part in it. (5) people and the use of other varieties such as carrots. Moreover, this practice has become more widespread in recent years, as the number of people practising it has increased by more than ten (10) in the space of ten years, and the varieties used have also increased. This is supported by Figure 3, which shows that the number of people practising rainfed and irrigated agriculture has risen from less than 5 30 years ago to almost 20 today, while those practising rainfed agriculture have remained constant.

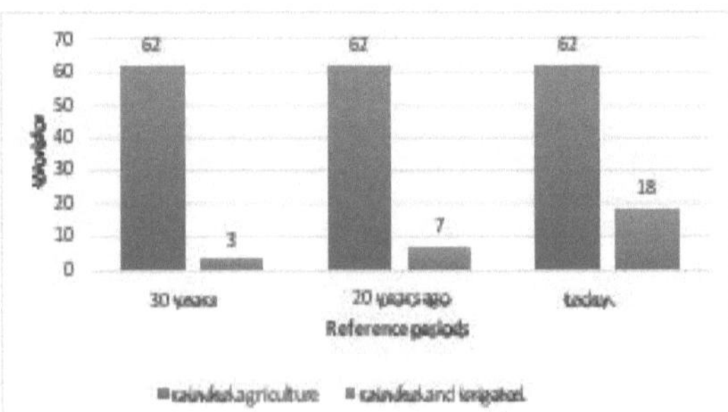

Figure 3: Types of farming practised in the Fandou Mayaki area, Source: field survey

2.2.1.1.2. Breeding

Livestock farming is the second largest sector of the rural economy. In recent years, like agriculture, livestock farming has also had to contend with constraints linked to the degradation of natural resources. Sedentary livestock farming has been the most common form of farming since the 1990s. 30 years ago, sedentary livestock farming accounted for over 95% of the sample surveyed. As for those practising both sedentary and transhumant livestock farming, 30 years ago they accounted for more than 32% of the sample, whereas today they account for less than 10% (Figure 4). This is linked not only to the degradation of resources, which has led to a fall in fodder production, and to poverty, which has prompted people to sell their animals, sparing only a few small ruminants, but also to insecurity, which hampers the mobility of people and their goods. This is shown in Figure 5, which shows that the areas where animals are reared are the plateaux and fields surrounding the villages, even though a minority of people say that their animals go far away. for grazing. This can also be justified by the statement made by the respondents that "In this locality, the most important animal species are goats, sheep and cattle. This is because these animals are the easiest to raise and sell to satisfy certain basic needs".

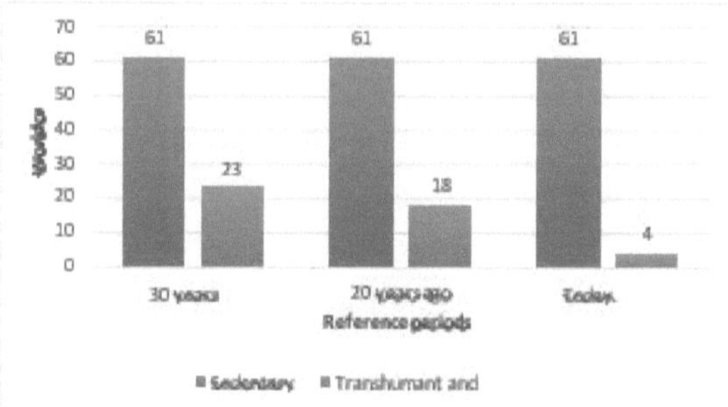

Figure 4: Types of livestock farming practised in the study area, Source: field survey

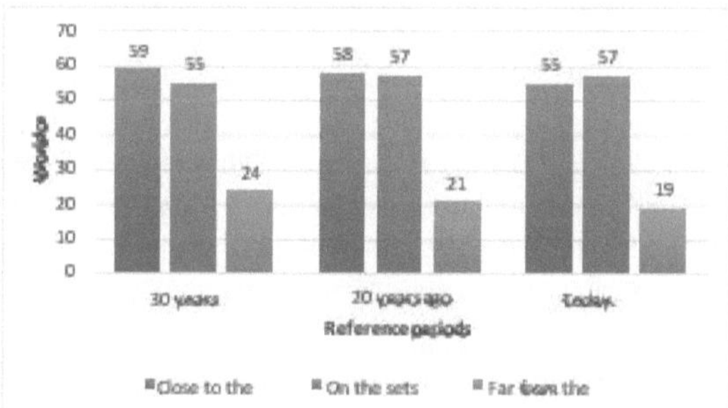

Figure 5: Grazing areas in the study area, Source: field survey

The by-products produced by the local population are mainly manure, milk, cheese and leather.

2.2.1.1.3. Hunting

Hunting is less common in this area. Figure 6 shows that of the 62 people surveyed, only one claimed to hunt, and that had been the case for 30 years.

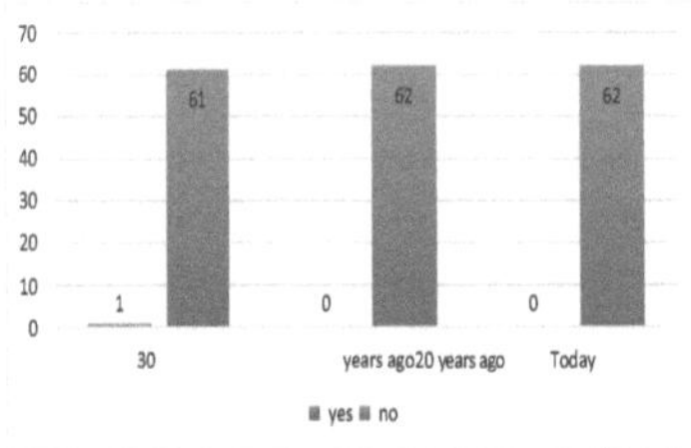

Figure 6: Hunting practices in the Fandou Mayaki area, Source: field survey

Nevertheless, according to the people interviewed, 30 years ago this area was home to several wild animal species such as gazelle, warthog, hare, squirrel, hyena, jackal, wild guinea fowl, partridge and wolf. However, most of these species have disappeared in recent years, according to the respondents, with the exception of a few such as partridge, hare, squirrel and wild guinea fowl, which are also rare. This is due to the disappearance of the plant cover that once served as a habitat for these animals.

2.2.1.1.4. Harvesting

Gathering is the third activity practised by this population as a service offered by nature. Gathering was once a very common practice in rural areas. It was the food source of our ancestors. But this activity is disappearing day by day. In this locality, only a few people claim to gather. The main plants picked by the local population are Annona senegalensis, Balanites aegyptiaca, Boscia senegalensis, Detarium microcarpum, Grewia bicolor, Sclerocarya birrera, Ziziphus mauritiana, Piliostigma reticulatum, Cadaba farinosa, Cassia tora and others. The main products harvested from these plants are fruit and leaves. But this practice is tending to disappear nowadays, given the gradual and uncontrolled extinction of certain plant species. According to the people interviewed, this is due to the recurrent droughts in the Sahel since the 1980s and the excessive cutting of wood for domestic purposes. This shows that harvesting locations vary from one period to another, as shown in Figure 7. It shows that those who gather far from the village are becoming more numerous, while those who gather close to the village are decreasing, even though the decrease is not significant.

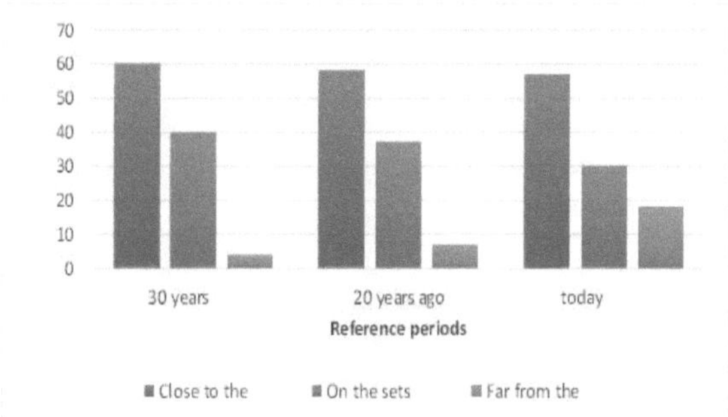

Figure 7: Picking locations by population in the area, Source: field survey

2.2.1.2. Medicinal resources

For medicinal resources, according to figure 8, 98% of the population surveyed said they practised sylvotherapy. Most of the practitioners claim to have practised this activity to treat illnesses such as haemorrhoids, abdominal pain or stomach aches, fever, ulcers and dysentery.

In the practice of sylvotherapy, the plant species most used by practitioners are : Annona senegalensis, Balanites aegyptiaca, Sclerocarya birrera, Ziziphus mauritiana, Piliostigma reticulatum, Bauhinia rufescens, Cassia sieberiana, Combretum glutinosum, Faidherbia albida, Khaya senegalensis, Prosopis africana and others. These plant species are a source of pharmaceutical products for people living in rural areas. However, according to the local population, this practice is being abandoned because of the disappearance of certain species and also because of advances in modern medicine.

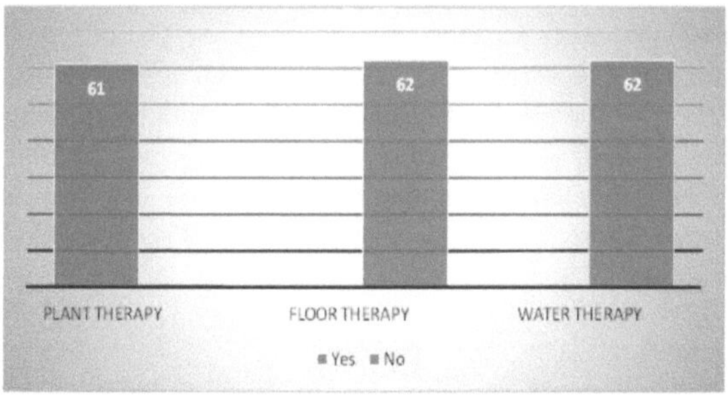

Figure 8: Use of resources for therapy, Source: field survey

As for the use of other resources such as water and soil for therapeutic treatment, all the people surveyed said they had not practised these activities.

2.2.1.3. Raw materials

2.2.1.3.1. Logging

In rural areas, wood plays a very important role and is used by almost the entire population. In this locality, it is used by 100% of the population surveyed. According to the respondents, this wood is first used for cooking (energy) and then for other activities such as building houses, sheds, granaries and fences for those who do not have the means to erect a wall. In this region, wood is used intensively because it is the main source of many rural activities.

A large proportion of the sample take their samples from the plateaux and surrounding areas. However, other people see the fields as their main harvesting areas. This can be explained by the fact that it is only on the plateaux that they manage to get wood to build sheds, granaries and fences, whereas in the fields they only manage to find wood for cooking.

2.2.1.3.2. Cultivation stems and harvesting areas

Crop stalks and wood are also used by the local population for a number of activities. These activities mainly involve building sheds, fences, showers, grain stores, etc. In recent years, they have also been used as fodder for cattle, due to the scarcity of other fodder species eaten by the animals. Others also harvest them for sale to meet certain social needs. The main areas where these stalks are harvested are the fields. This practice makes the fields vulnerable to any degradation process, as the stems help to combat wind and water erosion and also serve to trap certain elements that fertilise the soil.

2.2.1.3.3. Animal hides

In this locality, for the last 30 years, animal hides have been sold by most of the population. But there were a few people who used them as prayer rugs and shoes, but this is tending to disappear. This situation is justified by the fact that the local population mainly practises agriculture and considers that this type of activity is reserved exclusively for the Tuareg.

2.2.2. Cultural services

Cultural services are not included in the services provided by natural resources,nbut which are part of ecosystem services were investigated. Through these The different figures (Figure 9) show that 100% of the population surveyed replied that there were no places of worship in the Fandou Mayaki area. As for places of recreation, 60% said there were indeed places of recreation, compared with 40%. As for emblems, 92% confirmed that the population of this locality had no emblems, while 8% said that there were symbols by which the people of this locality were identified. These figures also show that cultural services, like other services, are losing their former value.

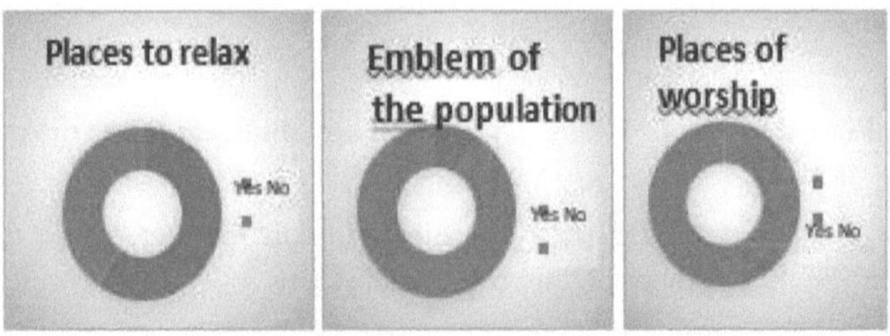

Figure 9: Cultural services identified in the locality of Fandou Mayaki, Source: field survey

2.2.3. Support services

The support or self-maintenance services identified in the study area are essentially limited to spaces preserved for animals and plants. These areas are of vital importance for maintaining biodiversity and preserving ecosystems. Unfortunately, however, as shown in figures 10 and 11, the majority of respondents said that these areas were no longer being conserved. In fact, for areas reserved for animals, less than 10 people answered yes that this type of area existed from 1990 to the present day, and these were located at the level of bodies of water and certain passageways that are now being nibbled away by farmers, or have even disappeared, whereas more than 80% said the opposite. As for spaces reserved for plants, between 1990 and 2000, fewer than 5 people said that these spaces existed, whereas today these spaces are no longer available in this locality (Figure 11).

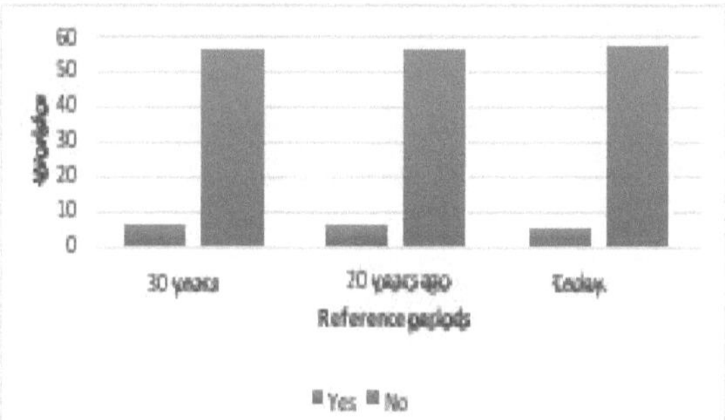

Figure 10: Spaces reserved for animals, source: field survey

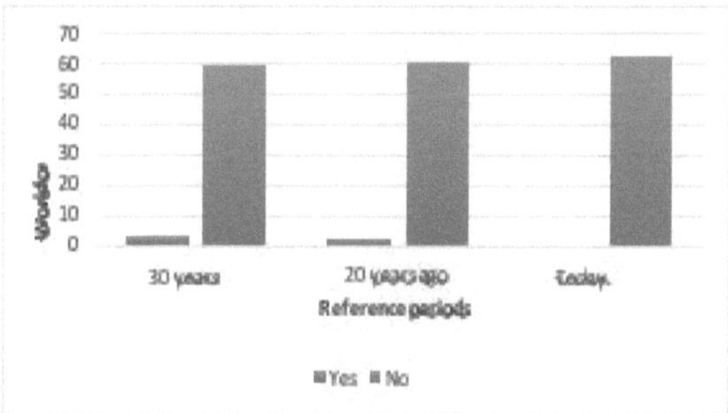

Figure 11: Spaces reserved for plants, source: site survey

CONCLUSION OF THE CHAPTER

In view of the above, this chapter shows that land degradation is progressing significantly in the Fandou Mayaki area. It varies from low to very high, with almost half the land area moderately degraded, while heavily to very heavily degraded areas also occupy a large proportion of the land area. This land degradation has had an impact on the provision of ecosystem services. As a result, an increasingly spectacular variation has been observed in access to and use of resources, and also in the practice of certain activities.

CHAPTER 3

BIOPHYSICAL AND SOCIO-DEMOGRAPHIC FACTORS OF LAND DEGRADATION AND ECOSYSTEM SERVICES IN THE FANDOU MAYAKI TERROIR

This chapter looks at the biophysical factors and then the socio-demographic factors of land degradation and ecosystem services through the analysis of climatic data, granulometric data, survey data and land use maps.

3.1. BIOPHYSICAL FACTORS

The biophysical factors that degrade land and ecosystem services are of several kinds, the most notable of which in the Sahelian region are the climate and the nature of the soils themselves. To this end, climatic and granulometric data specific to the study area are processed, analysed and interpreted.

3.1.1. Analysis of climatic data for the Filingué department

This section presents the climatic characteristics of the Filingué station by analysing parameters such as rainfall, temperature and wind. These parameters will make it possible to determine the effect of climate on natural resources and the ecosystem services they provide to mankind. However, the lack of data on temperature and wind means that we have to make do with the rainfall parameter alone.

3.1.1.1 Analysis of rainfall data from the Filingué station from 1988 to 2018

According to the World Meteorological Organisation, to determine the change in climate in a given environment, at least thirty (30) years of climate data on that environment are required. It is therefore important to focus on this recommendation when choosing the series of data to be processed, taking account of the data available. Recent data recorded by the Filingué Departmental Directorate of Agriculture (DDA/Filingué) at the Filingué station are available from 1988 to 2018. An analysis of this data shows that the rainfall pattern at this station is extremely variable. An analysis of the Lamb index results shows a significant difference between the high index of 2.14 mm in 1994 and the low index of less than 1.92 mm in 1993. Figure 12 shows that the period is characterised by alternating surplus and deficit years. This figure divides the series into three (3) periods: the first period (1988-1998) is marked by a spectacular fluctuation in precipitation, which is more or less constant from year to year. This can be seen in the succession of deficit and surplus years, with seven deficit years and three surplus years. Despite the peaks observed in this decade in 1994, we can deduce that it is the driest in the series, as the dry years far outnumber the wet ones. This situation therefore results in a rainfall deficit at this station during this decade. The second period (1999-2010) is characterised by a high variability of deficit and surplus years, with a predominance of surplus years. This suggests that this period is relatively wet compared with the previous one, as eight of the twelve years are wet, with a peak of 1.30 mm in 2001, the third wettest year in the series. The last period (2011-2018) is also marked by a succession of surplus and deficit years, with 5 surplus years out of 8. This is the wettest period of the episode, because as in the previous period, wet years far outnumber dry years. This period also includes the second wettest year in the series, 2016, after 1994. As a result, the years 1993, 2000, 2009 and 2011 are the most severe deficit years of the episode. Whereas 1994, 2001 and 2016 were the years with the highest surpluses. The graph also shows that the rainfall trend in this series is upwards.

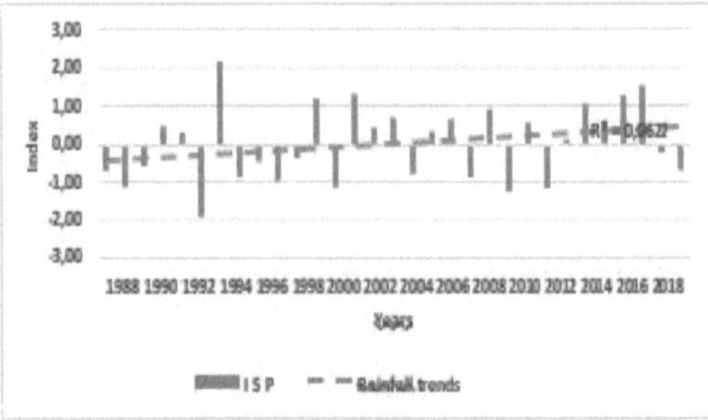

Figure 12: Changes in rainfall at the Filingué station from 1988 to 2018 according to Lamb anomaly index, Source: **DDA/ Filingué**

The rainfall index shows a high degree of variability in the rainfall regime in Filingué district, despite an upturn in rainfall from 2005 onwards, as shown in Figure 13. This inter-annual variation in rainfall has direct consequences for the water balance and also accentuates certain climatic risks observed in the department. Figure 13 also shows that there is an inter-annual variation in rainy days at this station. The graph shows that 1994, 2003 and 2010 were the years in which several rainy days were recorded. Admittedly, these were surplus years in terms of rainfall, but they do not justify the rainfall variations observed in the index. The wettest year, 1994, only recorded 35 rainy days, while 2003, which recorded the highest number of rainy days (36), is not even ranked among the top five wet years. In contrast to previous years, this figure shows a downward trend in the number of days, while rainfall in the area has picked up. This suggests that the increase in rainfall is not due to the number of rainy days, but to the increase in rainfall intensity. And this has indirect repercussions on land degradation through water erosion.

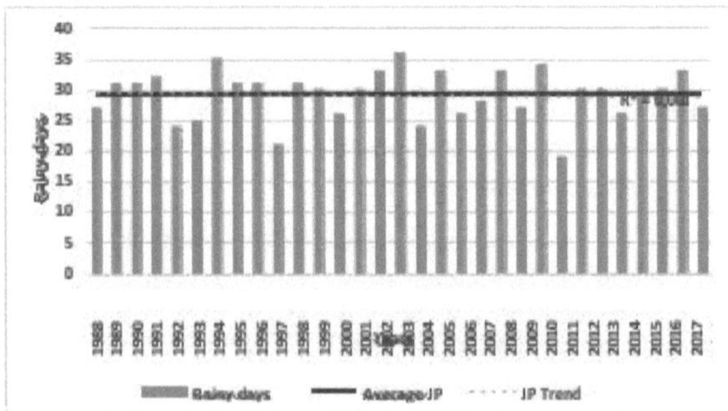

Figure 13: Number of rainy days per year at Filingué station from 1988 to 2018, Source: DDA/ Filingué

Despite the lack of data on other climatic parameters, analysis of the rainfall data for this department suggests that the climate is subject to increasingly recurrent variations, often leading to a resurgence of certain phenomena. According to several respondents, recurrent droughts, especially during the 1980s, led to the disappearance of vegetation on the plateaux.

The disappearance of plant cover has led to the emergence of certain climatic risks, which have brought about social and environmental changes. These risks can be summed up as droughts during the rainy season, and sometimes floods, linked either to insufficient or unavailable rainfall. In recent years, these risks have led to water and wind erosion, which has had a negative impact on the locality's agricultural and fodder production. Figure 15 clearly illustrates this idea, as all 62 people surveyed replied that the climatic risks they observe today are erosion and flooding, and 58 people said that drought during the rainy season and the late start of the rainy season are the risks they face today. Whereas 30 years ago, in the 1990s, the climate risks identified in this locality were excessive rainfall, extreme cold and high heat.

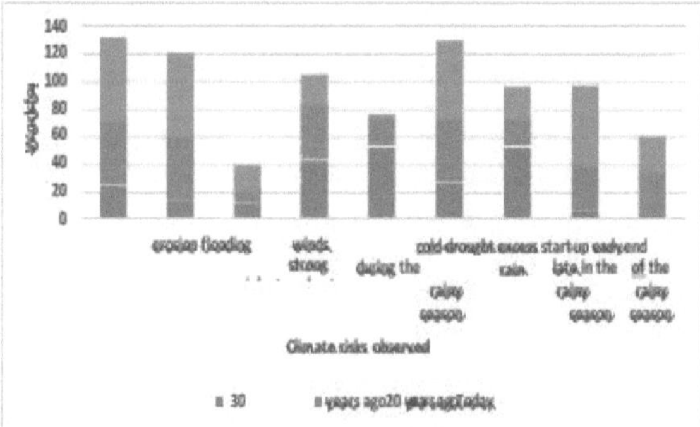

Figure 14: Climate risks observed by farmers in the Fandou Mayaki area from 1990 to 2020, Source: field survey

These risks are the cause of the deterioration of the land and the living conditions of the local population. As shown in Figure 15, out of a sample of 62 respondents, more than 50 said that climate hazards have serious effects on the soil, which they destroy, as well as on agricultural and fodder production.These risks also lead to illnesses that compromise the lives of a population that is already living in poverty. weakened.

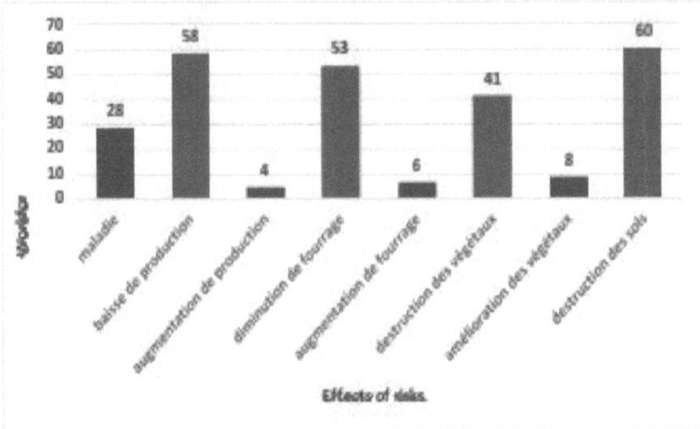

Figure 15: Effects of climate hazards on the local population and environment de Fandou Mayaki, Source: field survey

Land degradation can also be justified by analysing granulometric data, which will determine the texture of the soils in this locality and their susceptibility to erosion.

3.1.2. Textural characteristics of soils in the Fandou Mayaki terroir Granulometric analysis is used to quantitatively determine the distribution of soil particles by diameter class. It involves the interpretation of histograms and curves.

cumulative frequencies constructed from the rejects retained by the various sieves and the percentages of the granulometric classes. Finally, it enables the soils of the different geomorphological units to be classified in terms of their textural composition and their behaviour in the face of erosive agents. To do this, the median was determined from the cumulative curves and granulometric indices (grain classification and asymmetry index) were calculated according to the formulas of P. TRASK, (1932), R. FOLK and W. WARD, (1957) on the basis of quartiles and percentiles. The median represents the size corresponding to 50% of the grains determined directly on the cumulative curve; the classification index (So) provides information on the hydrodynamics during sediment deposition. It also indicates the quality of the classification; the lower it is, the better the sediment is classified and the asymmetry index (Ski) reflects the degree of asymmetry of the distribution curve in relation to the median. This parameter provides information on enrichment in coarse particles (negative asymmetry), enrichment in poorly sorted fine particles (positive asymmetry), or whether there is symmetry (M. TOSSOU EDJROGAN et al. 2019). However, by referring to the granulometric classification of K. WENTWORTH, (1922), it emerges from this analysis that, the samples of sands taken in the terroir of Fandou Mayaki are made up of six classes: very coarse sand; coarse sand; medium sand; fine sand; very fine sand and silt. The particle size composition of these classes differs from one unit to another.

3.1.2.1. The characteristics of the soils on the slopes of the plateaux

Of these units, only one sample was taken from the tiger bush on the southern plateau. The other areas were not sampled because of their rocky texture, as shown in photo 6.

Photo 6: Bare slopes of the Kouboutché (A) and Gorma (B) plateaus

Analysis of the sample taken from the reverse side of the southern plateau (T2) shows that its particle size class is essentially made up of coarse particles, which account for 61.40% of the sample, with a predominance of very fine sand (17.90%) and very coarse sand (12.80%). Very fine particles accounted for 38.60%. This large quantity of coarse particles compared with very fine particles shows that the soils in this unit are less valuable from the point of view of plant and crop production.

The relative frequency histogram for this sample (Figure 16) shows that it is unimodal at the 0.063 mm sieve, which explains why the particle deposition process is carried out by a single transport agent. The cumulative frequency curve is parabolic, with a spectacular rise from 8% to over 60%.

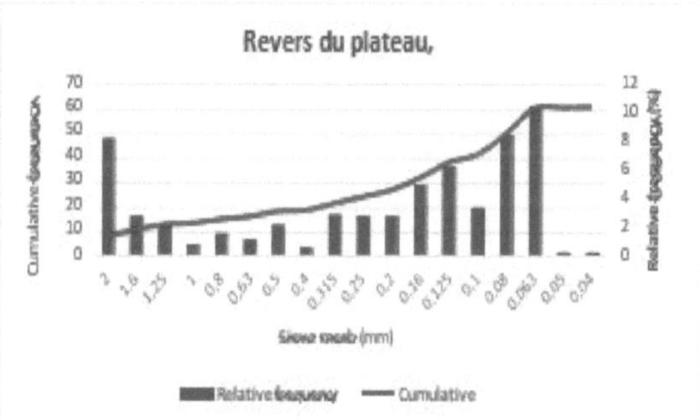

Figure 16: Cumulative and relative frequency curves for soils on the reverse side of the southern plateau

3.1.2.2. The characteristics of piedmont soils

Two samples were taken from the foothills. The soil characteristics of these samples were as follows: Analysis of the soils taken from the foothills (T1) reveals that 54.2% are made up of very fine particles smaller than 0.04 mm and coarse particles between 0.04 mm and 2 mm represent 45.8%, with a predominance of fine and medium sands, which account for 17.20% and 14.60% of the coarse particles. Fine particles accounted for 7.60% of the sample. This predominance of very fine particles compared to coarse particles suggests that the quality of the soil at this location is appreciable. The relative frequency histogram for this sample (Figure 17) shows two modes, the primary equal to 0.8 mm and the secondary 0.63 mm, which explains why there are two types of deposit from different agents.

The particle size composition of the second foothills (T2) is similar to that of the reverse side of the plateau, as it is also predominantly made up of coarse fractions, which occupy 92.40% of the sample. They are dominated by fine sand (29.60%) and very fine sand (29%). On the other hand, very fine particles are less important, representing only 7.60% of the entire sample.

This implies that this unit has the same appreciable characteristics as the unit on the reverse side of the plateau. A reading of its relative frequency histogram (Figure 17) shows that it is multi-modal, with the main mode observed at the 0.08 mm sieve. This suggests that sediment is deposited at this level either by a single transport agent acting in several forms or by several agents acting differently.

The cumulative frequency curves for these units have the same shape as the reverse plateau unit. They are characterised by a parabolic shape ranging from 0.1% to 45.2% for the piedmont1 curve and from 1.3% to 91.6% for the second piedmont.

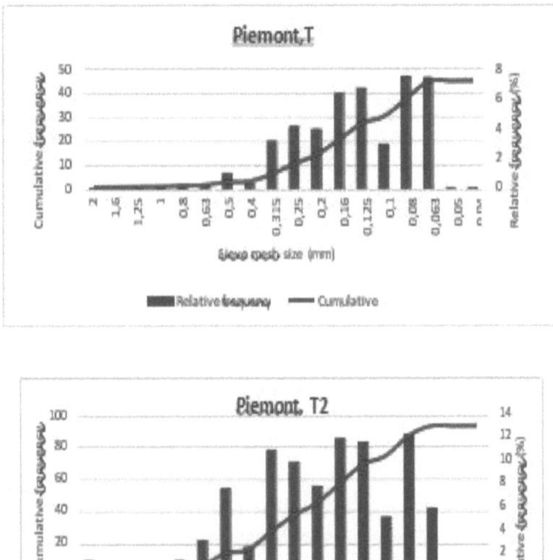

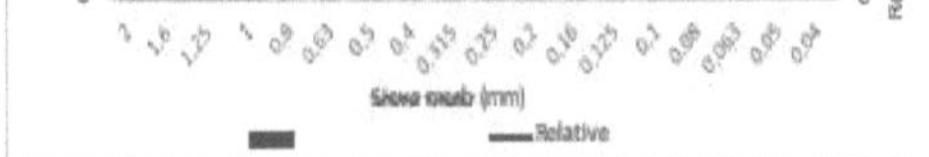

Figure 17: Cumulative and relative frequency curves for soils sampled in the foothills units

3.1.2.3. The characteristics of glacis soils

Four samples were taken from the glacis units, two from the glacis in transect 1 and two from the glacis in transect 2. The particle size data for the soil sample from the upstream glacis of transect 1 show that coarse particles greater than 0.04 mm dominate over very fine particles less than 0.04 mm. These coarse particles represent 88.7% of the sample, the most dominant particles being very fine sand (49.60%), fine sand (18.30%) and silt (11.30%). Very fine particles accounted for 11.30% of the sample. The predominance of coarse particles at the of this unit suggests that its soils are less fertile than those of the previous unit. Figure 18 also shows that the relative frequency histogram for this sample is unimodal at the 0.08 mm sieve. This implies that the deposition process is caused by a single transport agent acting exponentially. The cumulative frequency curve for this unit varies from 0% to 87.7% with a degraded parabolic shape. For the downstream glacis unit, the sample taken at this location is mainly composed of coarse particles, which occupy 94.4% of the sample, with a predominance of fine sand (34.6%) and medium sand (30.20%). The fine particles in this sample are less significant at only 3.70%. As for the very fine elements, they occupy only 5.6% of the whole sample. This exorbitant discrepancy between the textural composition of this sample shows that it is very sandy and, like the previous sample, has a less appreciable character. The relative frequency histogram in Figure 18 indicates that in this unit (downstream glacis) the deposition process is carried out by a single transport agent, since it is also unimodal like the previous unit, whose mode is observed at the 0.315 mm sieve. Its cumulative frequency curve with its convex shape ranges from 0.2% to 92.9%.

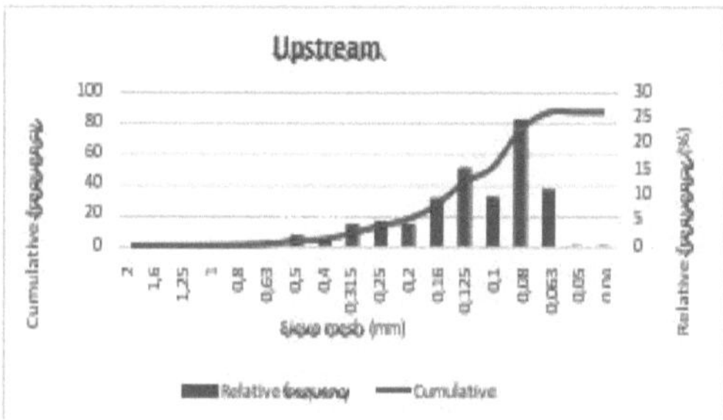

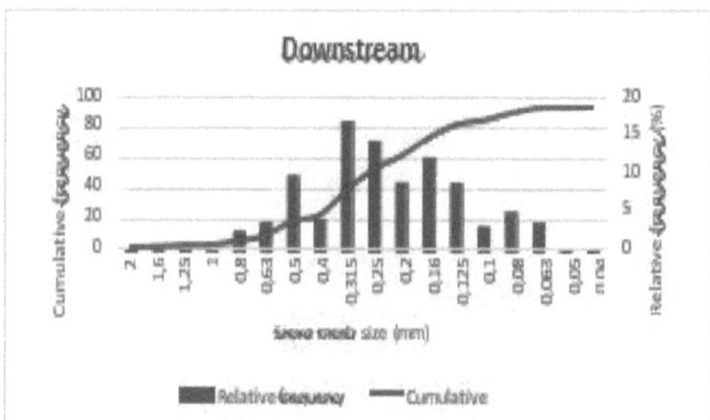

Figure 18: Cumulative and relative frequency curves for soils sampled on the glacis of transect 1

The upstream and downstream glacis of transect 2 constitute the second sub-unit of the glacis unit. The textural composition of the samples analysed at these levels is predominantly made up of moderately coarse particles including very fine sands (34.10%) and medium sands (20%) for the upstream glacis and 30% fine sands and 26.70% very fine sands for the downstream glacis. Fine clay and silt particles represent 8.90% for the upstream part and 7.20% for the downstream part, while very fine particles occupy 8.8% and 10.9% of the samples. This grain structure means that the particle size class of this unit is essentially sandy. The histograms of the relative frequencies of these sub-units (Figure 19) show that they each have a single mode, which can be seen at the 0.08mm sieve for the upstream glacis, and that of the downstream glacis can be seen at the 0.315mm diameter. The cumulative curves show the same shape, with a frequency variation ranging from 0.1% to 91.3% for the upstream glacis and from 0.1% to 88.6% for the downstream glacis.

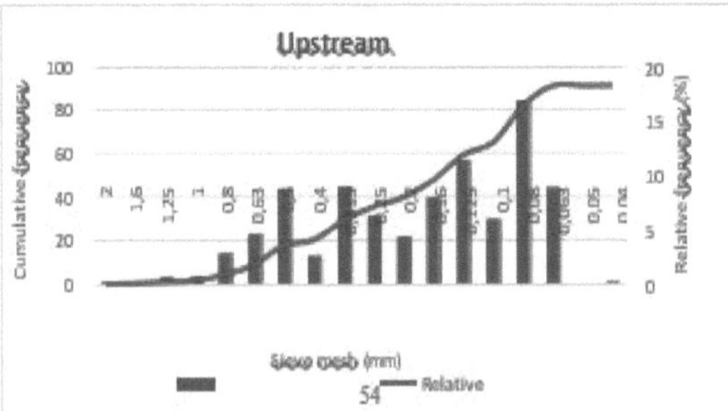

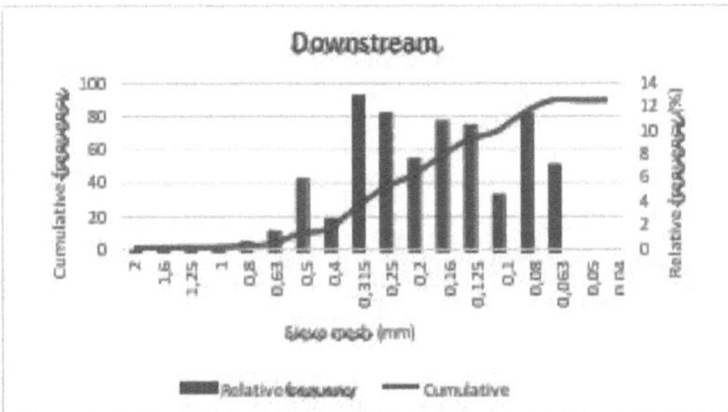

Figure 19: Cumulative and relative frequency curves for soils sampled on the glacis of transect 2

3.1.2.4. The characteristics of valley floor soils

Valley bottom units 1 and 2 have almost the same granulometric characteristics. This is due to the fact that they occupy the same geomorphological unit and also to their almost identical degree of use. The most dominant textural class in these samples (valley bottom 1 and 2) is that of coarse particles, which represent 84.8% for the first sample and 88.3% for the second. These coarse sands are made up of very fine sands (32.30%) and fine sands (22.90%). 42%

for the valley floor sample1 and 18.30% for the valley floor sample2. Fine fractions (silt) play a significant role, with 11.60% for the valley floor1 and 13.30% for the 2ème valley floor. Very fine particles also account for a significant proportion, at 15.2% and 11.7% respectively. A reading of the figures for this unit (Figure 20) shows that the histograms of relative frequencies for these units also show a mode each whose modal class is observed at the 0.08 mm sieve level for both samples. This suggests that these two units are affected by a single transport agent depositing sediment at these locations. The cumulative curves, as for the glacis unit, show the same shape with frequencies varying from 0.1% to 83% for the valley floor1 and from 0.1% to 87.1%.

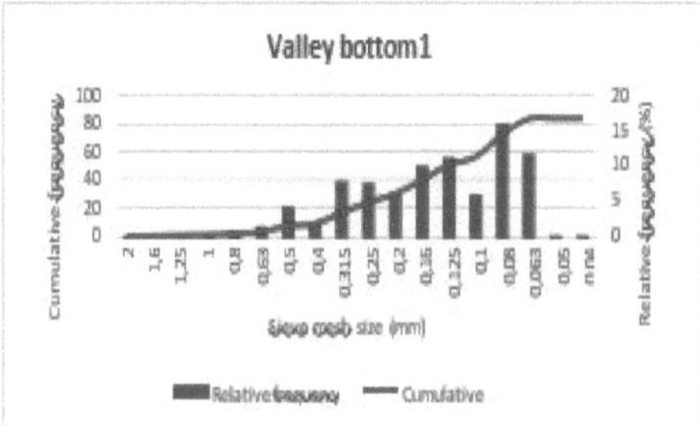

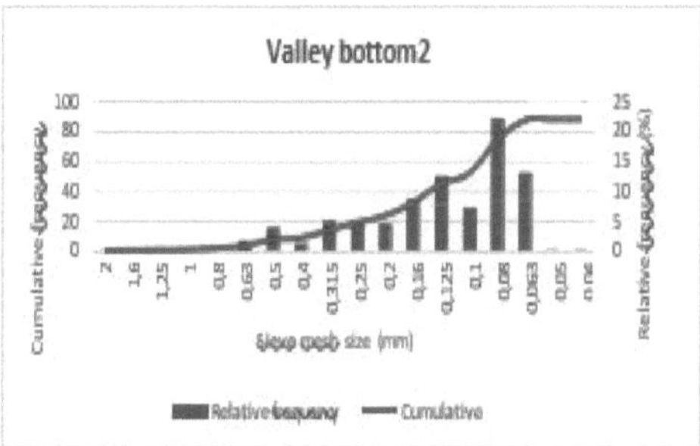

Figure 20: Cumulative and relative frequency curves for soils sampled on the valley floor

3.1.2.5. Characteristics of lowland soils

The lowland unit represents the last unit observed. The soils in this unit share the same textural characteristics as those in the glacis. Its granular composition is made up of fine and medium sands, which represent 36.40% and 35.40% of the sample taken. Fine silt particles are less significant in this area, at 2.40%, as are very fine particles, which only represent 5%. This granular classification leads us to conclude that this unit is composed mainly of sand. This gives it a less appreciable character. The relative frequency histogram for this unit (Figure 21) shows that the 0.315 mm sieve represents the only modal class identified at this level. This shows that it is the only sediment transport and deposition agent on this unit. The cumulative curve has the same shape as that of the transect's downstream glacis. It varies from 0% at the 2mm sieve to 93.9% at the 0.04mm sieve.

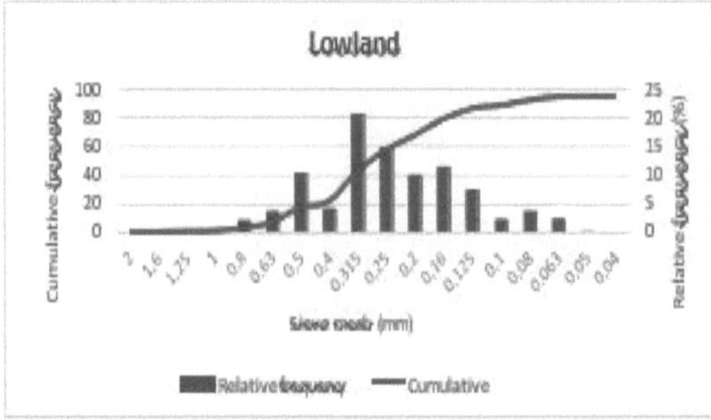

Figure 21: Cumulative and relative frequency curves for soils sampled in the lowlands

Analysis of these different samples shows that the cumulative frequency curve on the reverse side of the plateau has a related linear shape, whereas those for the other units are parabolic and degraded. This difference in the shape of the curves suggests that the deposition process is characterised by heterogeneity in the transport agents between these different units. Thus, the related linear shape reflects sediments that are in the process of being formed by the disintegration of rocks, and that have not undergone any transport. It also corresponds to multimodal histograms and cumulative curves of linear or related facies. The degraded parabolic shape indicates a pronounced inflection on the side of the coarse fractions and a symmetrical inflection on the side of the fine ends. This reflects the presence of materials with unimodal and bimodal distributions. In addition, analysis of the physical composition of soil samples from these landscape units shows that these soils have particular characteristics in relation to each other and act differently in relation to erosion, depending on their textural composition. In fact, the results of this analysis show that the piedmont1 sub-unit has a very appreciable soil textural composition, since it is composed mainly of fine particles. This high proportion of fine elements means that the soil in this unit is stable but sensitive to erosion. It also makes the soil structure very compact, thereby reducing the infiltration that encourages plant regeneration. This situation leads to a loss of vegetation, which protects the soil against certain risks and also traps its elements when erosive agents act. This is confirmed by what the respondents said: "The plant formations that existed in this locality 30 years ago fought against climatic risks such as erosion, flooding and violent winds, even though they were shaken by droughts. The aforementioned risks were not experienced in this area at all 30 years ago, but in recent years, they are the most threatening risks". This shows that the plant formations present today are no longer able to counteract these risks, which are becoming more and more recurrent. Nevertheless, they continue to play an important role in mitigating violent winds and high temperatures. In contrast, the other units are characterised by a predominance of coarse particles, which account for more than 80% of the samples. This considerable presence of coarse particles indicates that the soils in these units have a macroporous structure, which greatly favours the infiltration of water and the development of plants, thereby reducing the force of erosive agents. However, they remain sensitive to mechanical erosion caused by the use of farm implements. Photo 7 shows the two situations described above.

Photo 7: Arable field threatened by encrustation (A) and fallow land replanted with formation of nebkas (B)

In addition, analysis of the physical composition of the soil samples shows that these soils have almost similar textural characteristics to one another, as they are predominantly made up of more than 50% sand, except on the reverse side of the plateau and in the foothills1 where fine particles dominate the sample. As mentioned above, this predominance of sandy particles favours water infiltration, making it easier for plants to germinate and crops to withstand pockets of drought during the rainy season.

Despite this, the results of calculating the indices, as shown in Table 6, show that the soils in these units can act differently in the face of erosive agents.

Table 6: Summary of soil characteristics sampled in the Fandou Mayaki terroir units, Source: our study

Units	Modes	Ski	So	Comments
Reverse of the tray	0,063	0,8	0,3	A single transport agent: wind. Particles are well classified and highly asymmetric towards fine particles.
Piedmont	0,08 ; 0,63	0,8	0,3	Two transport agents: water and wind, particles are well classified and highly asymmetric towards fine particles
Upstream glacis	0,08	0,8	0,3	A transport agent: water, particles are well classified and very asymmetric towards fine particles
Downstream glacis	0,08 ; 0,315	0,8	0,32 ; 0,3	Two transport agents: water and wind, particles are well classified and highly asymmetric towards fine particles
Valley bottom	0,08	0,8	0,3	A transport agent: water, particles are well classified and highly asymmetric towards fine particles
Lowland	0,315	0,8	0,3	A transport agent: water, particles are well classified and highly asymmetric towards fine particles

This hypothesis was confirmed by field observation, which showed that the most dominant factors were water and wind erosion, as shown in photo 8.

A B

C

D

A: bank undercutting at foothill level; **B**: gully erosion at foothill level; **C**: arable field threatened by deflation at downstream glacis level; **D**: fallow land threatened by deflation at upstream glacis level. **Photo 8:** Types of degradation identified on the different geomorphological units

A comparative analysis of the above granulometric analysis shows that all the granulometric curves show a predominance of fractions between 0.063 mm and 1 mm. These fractions are made up of very fine sands, fine sands and medium sands. In addition, the various indices calculated show that these soils have similar characteristics in terms of granular classification and asymmetry. As for the modal classes determined, they show that the majority of the samples are unimodal and have almost all have the same modal classes, i.e. they are all confronted with the same phenomenon of sediment transport and deposition. These various similarities suggest that the soils sampled from the different units in particular, and those of the terroir in general, face the same types of degradation caused by the same factor, namely erosion. This erosion is linked not only to climatic variability but also to human activity.

3.2. SOCIO-DEMOGRAPHIC FACTORS

For several decades, galloping population growth has been classified by many researchers as a factor in the degradation of natural resources. However, population growth is not in itself a factor in the deterioration of resources, but it is the use that the population makes of the resource that determines whether or not it is degrading. In this section, we will first demonstrate the land use and management methods adopted by the local population, and then analyse the consequences of this use on the various land use units by analysing the land use maps.

3.2.1. Land use and management in the terroir of Fandou Mayaki

To meet their basic needs, the local population has developed activities such as farming and livestock rearing. These activities were once so promising that they were practised intensively. But for some time now, the produce from these activities has been falling exorbitantly, and no longer yields what it used to. To make up for this shortfall, local people have taken corrective action by extending the area under cultivation, leading to uncontrolled clearing of the land and the adoption of other forms of livestock farming. The implementation of these new measures has led to increasing over-exploitation of the land, resulting in a decline in fertility. This in turn has led to a fall in agricultural and fodder production. Table 7 shows the agricultural production situation from 1990 to 2020.

Table 7: Farmers' perceptions of agricultural production from 1990 to the present, Source: field survey

Production (number of bundles)	Workforce 30 years ago	Workforce 20 years ago	Workforce today	Frequencies 30 years ago	Frequencies 20 years ago	Frequencies today
No answer	1	1	0	1,60%	1,60%	0,00%
Less 200 or more	24	38	51	38,70%	61,29%	82,30%
200 à 400	14	13	11	22,60%	20,98%	17,70%
400 à 600	8	6	0	12,90%	9,68%	0,00%
600 and more	15	4	0	24,20%	6,45%	0,00%
Total	62	62	62	100%	100%	100%

This table shows that the populations of this area are experiencing a drastic drop in agricultural production. By converting the number of bunches into kilograms, i.e. 10 bunches to 100kg, we can see that the number of people producing less than 200 bunches, i.e. 2000kg, which corresponds to 2 tonnes, has only increased from 30 years ago to the present day, rising from 24 to 51 people, while the number of other classes has only decreased. We also note that 80.30% of those surveyed have harvested fewer than 200 bunches, i.e. 2 tonnes, over the last two years, compared with 38.70% 30 years ago. On the other hand, those producing more than 600 bunches (6 tonnes) have gone from 24.20% 30 years ago to 0% today. As for the decline in fodder production, this is also justified by the opinions of the respondents, who admit that "30 years ago, livestock feed in this locality consisted mainly of straw, fodder from the air and bran. But in recent years, other feed supplements have been added. These include crop residues (stalks, bean residues, etc.), which are collected and fed to the animals as soon as the harvest is finished. This in turn leads to land degradation, as it deprives the soil of the elements it needs not only to improve organic matter but also to protect it from erosion. Faced with this situation, farmers have found themselves obliged to adopt other alternatives. These strategies consist of adding organic manure and household waste to the fields, using chemical fertilisers or, for those who have the means, keeping livestock in the fields as a contract.

3.2.2. Analysis of land use dynamics in the terroir of Fandou Mayaki

The study of land use dynamics in this area was based on the analysis of satellite images over a 30-year period. Maps for different periods were drawn up, covering 20 years and 10 years, i.e. 1990, 2010 and 2020. These maps show the changes taking place in the area.

3.2.2.1. Change in land use from 1990 to 2020

The interaction between man and nature has meant that from 1990 to 2020, i.e. 30 years, a certain amount of change has been revealed in the land-use classes of the Fandou terroir. These changes can be seen in Table 8.

Table 8: Surface areas ofland use units in 1990, 2010 and 2020, Source: our study

Land use classes	Surface area (ha)			Surface area (%)		
	1990	2010	2020	1990	2010	2020
Degraded scrub	57,6	15,03	58,41	0,37	0,10	0,38
Degraded tiger bushes	278,82	22,5	56,43	1,81	0,15	0,37
Regular tabby bushes	195,84	-	-	1,27	-	-1,27
Degraded riparian degraded riparian formations	208,35	123,93	100,53	1,35	0,80	0,65
Formations regular riparian training	65,61	89,55	13,14	0,43	0,58	0,09
Human settlements	66,69	162,63	255,87	0,43	1,06	1,66
Fallow land	1665,72	672,12	1003,95	10,82	4,36	6,52
Lake	-	33,39	225,45	-	0,22	1,46
Bare surfaces	1324,17	2017,8	1112,49	8,6	13,10	7,22
Rocky ground	1409,31	1479,06	1489,14	9,15	9,60	9,67
Growing areas	10127,79	10783,89	11084,49	65,77	70,03	71,98
TOTAL	15399,9	15399,9	15399,9	100	100	100

This table shows that the land use units in the study area have varied in size over the 30 years.

In fact, the surface areas of certain units have gradually changed. These include cultivated areas. These, which are the largest units in terms of surface area, have seen their surface area increase from 1990 to 2020, rising from 10127.79 ha in 1990 to 10783.89 ha in 2010 and reaching 11084.49 ha in 2020, an increase of 956.7 ha. After the cultivated areas come the rocky areas, which are also units that have seen a gradual increase in their surface area throughout the series. The proportions of these units are estimated at 1409.31 ha in 1990, 1479.06 ha in 2010 and 1489.14 ha in 2020. In addition to cultivated areas and rocky terrain, human settlements and bodies of water have also seen their surface area evolve.during this period. Human settlements, which were 66.69 ha in 1990, have risen to 162.63 ha in 2010 and will increase to 255.87 ha in 2020. The same applies to water bodies. The latter, which did not even exist in 1990, made their appearance in 2010, when their extension is estimated at 33.3 ha, rising to 225.45 ha in 2020, representing an increase of more than 192 ha. In contrast to the previous units, the surface areas of the other units have varied both progressively and regressively. Fallow land, which is second only to arable land, has seen a remarkable reduction in area, especially between 1990 and 2010. In 1990, fallow land was estimated at 1,665.72 ha, whereas by 2010 it had shrunk to 672.12 ha, a reduction of 993.6 ha. However, between 2010 and 2020, the surface area of these units increased, but less significantly than in the previous situation, rising from 672.12 ha to 1003.95 ha, a change of 331.83 ha. The surface area of bare land was estimated at 1324.17 ha in 1990, whereas in 2010 this same surface area was estimated at 2017.8 ha, i.e. an increase of 693.63 ha, in contrast to 2020 when it fell further to 1112.49 ha, i.e. a decrease of 905.31 ha between 2010 and 2020. Like fallow land and bare areas, the other units (degraded scrub, degraded tiger scrub and regular riparian formations) are the ones that have changed the most during this period. The surface area of the first two decreased between 1990 and 2010, but showed a slight increase between 2010 and 2020, whereas the opposite was true of the regular riparian formations. Degraded riparian forest formations and regular tiger scrub are in the opposite situation. Degraded riparian forest, which in 1990 was 208.35 ha, is estimated to be 100.53 ha in 2020, a drop of more than 100 ha. Regular tiger scrub, which existed in 1990, has completely disappeared from the sector between 2010 and 2020. This analysis reveals that of the 11 land use units identified in the study area, 4 units have seen their surface area gradually increase. While 5 units have experienced both a progressive and regressive change in their surface area. However, the situation of two other remaining units is more critical, as one has seen its area diminish in a regressive manner and the other has completely disappeared to make way for other units.

This can be clearly seen on maps 6 and 7.

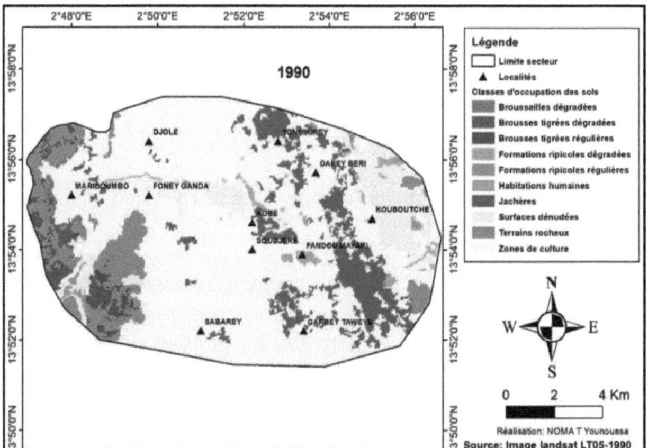

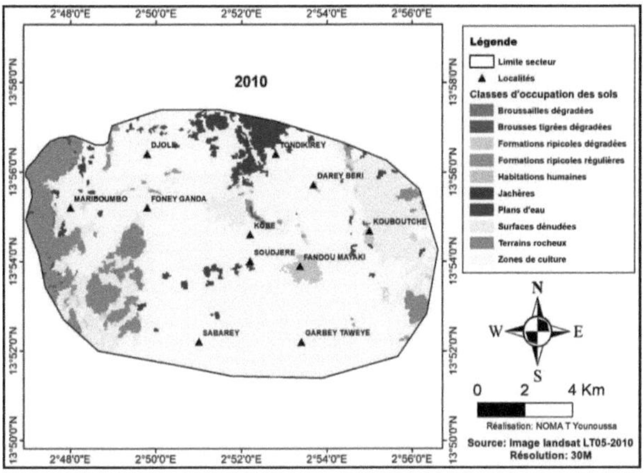

Map 6: Land use in the Fandou Mayaki area in 1990 and 2010

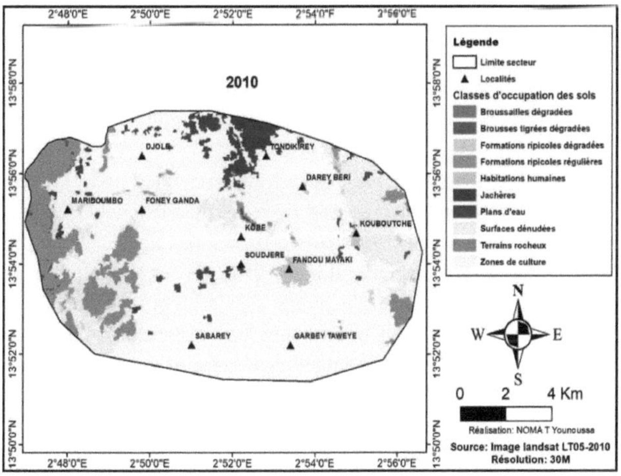

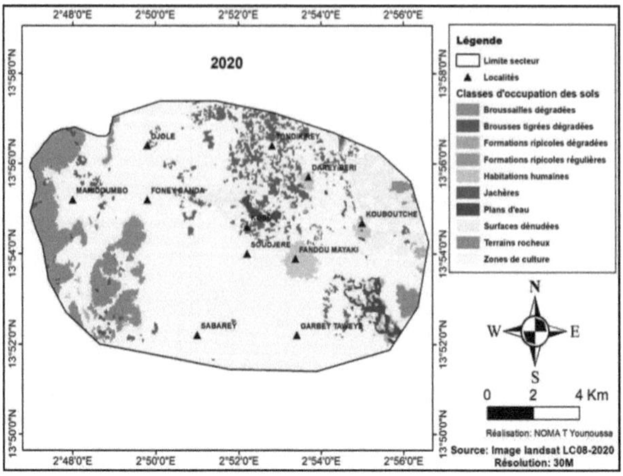

Map 7: Land use in the Fandou Mayaki area in 2010 and 2020

3.2.2.2. Changes in land use between 1990 and 2020 Analysis of the various maps shows that the study area has undergone a remarkable dynamic in land use and occupation over these 3 decades. These changes have mainly affected cultivated areas, bodies of water, fallow land and regular tiger scrub. As shown in 8ème

Between 1990 and 2020, the surface area of these units declined and expanded significantly. The surface area of cultivated areas has spread by more than 6%, and that of water bodies by more than 1%, compared with 0% in 1990. However, regular tiger bush and fallow land have shrunk by more than 1% for the former and more than 4% for the latter. The area of human habitation also increased by more than 1%, while that of degraded tiger bush and bare land also fell by more than 1%. The changes in the other land-use units are less significant, but are not negligible, as they could lead to irreversible degradation if they continue at the same rate.

Table 9: Changes in land use between 1990 and 2020, source: our study

Land use classes	Surface area (%)		Changes (%)
	1990	2020	1990-2020
Degraded scrub	0,37	0,38	+0,01
Degraded tiger bushes	1,81	0,37	-1,44
Regular tabby bushes	1,27	0	-1,27
Degraded riparian formations	1,35	0,65	-0,7
Regular riparian formations	0,43	0,09	-0,34
Human settlements	0,43	1,66	+1,23
Fallow land	10,82	6,52	-4,3
Lake	0	1,46	+1,46
Bare surfaces	8,6	7,22	-1,38
Rocky ground	9,15	9,67	+0,52
Growing areas	65,77	71,98	+6,21

Figure 22 shows the dynamics of land use units in this locality from 1990 to 2020.

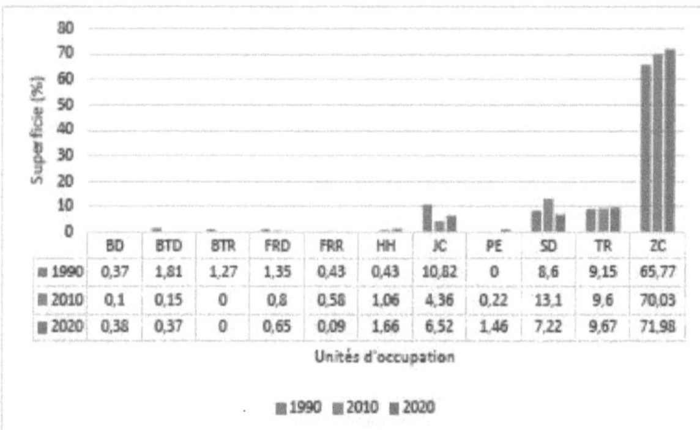

Legend: BD: degraded scrub, **BTD:** degraded tiger scrub, **BTR:** regular tiger scrub, **FRD:** degraded riparian formations, **FRR:** regular riparian formations, **HH:** human habitations, **JC:** fallow land, **PE:** water bodies, **SD:** bare surfaces, **TR:** rocky terrain, **ZC:** cultivated areas.

Figure 22: Change in land use units between 1990 and 2020

CONCLUSION OF THE CHAPTER

In short, this chapter shows that the factors leading to land degradation and the degradation of ecosystem services in the terroir are of several types. These are determined by climatic parameters through their frequent variations, the physical characteristics of the soil through its ability to resist erosion, and the behaviour of humans through their various land occupation, use and management activities. This analysis has shown that these parameters act collectively to bring about a change in the environment and the natural resources it contains.

CHAPTER 4
DISCUSSION OF RESULTS AND DESIGN PROPOSAL

This final chapter first discusses the results of this research, and then makes a number of proposals for the future. layout alternatives

This research was carried out using several methods: field observation, mapping (satellite image processing), analysis and interpretation of survey data, climate data and granulometric data.

4.1. DISCUSSION OF RESULTS

4.1.1. Biophysical factors of land degradation

Analysis of the climatic and granulometric data reveals that the biophysical factors of land degradation in this zone result not only from climatic variability but also from soil texture. The results of the climatic data analysis confirm this hypothesis. It emerges from this analysis that the rainfall parameter in this locality is characterised by increasingly recurrent oscillations between wet and dry years. These rainfall fluctuations sometimes result in heavy rainfall, sometimes in drought, which has an impact on water and wind erosion, transporting soil nutrients. This stripping of materials exposes the soil to vulnerability, compromising its various functions. Generally speaking, this phenomenon can be observed throughout the African continent. This is confirmed by J E. PATUREL et al, (1998) who state that the annual rainfall series recorded at a number of stations covering non-Sahelian West and Central Africa show that dry and wet periods alternate in this region. This is also the case in the Sahel and has been demonstrated by several researchers (OZER P. et al : 2010 ; MAJOLAINE C.et al : 2012 ; F. ROGER et al : 2011) reported by B. BAGAGI ARZIKA, (2017). These authors have shown that the Sahel has experienced a significant drop in rainfall over recent decades. According to analyses by the United Nations Environment Programme, nineteen climate "hotspots" have been identified in West Africa (UNEP, 2011) reported by F. GEMENNE et al. (2017). These are areas that have been hit hard by flooding in recent years, but are also experiencing rising temperatures and a substantial increase in the frequency of droughts.

This proves that floods, droughts, disruption of rainy seasons and heat waves are the most tangible extreme weather events that affect people.West Africa, as demonstrated by DARA (2013), quoted by F. GEMENNE et al. (2017). This significant rainfall variation is also highlighted in particular by P. OZER et al. (2015), who show that Niger has also been facing profound environmental, social and economic changes in recent decades, often linked to increasing rainfall stress.

In addition, the granulometric composition of the soils in the area, with a predominance of sandy and silty fractions, means that these soils are exposed to all kinds of erosive hazards. In fact, these soils are vulnerable not only to water erosion, which, through heavy downpours, removes sand clods by splashing, transporting them away from the site, but also to wind erosion, which transports large quantities of fine particles. The impact of these erosive processes is gullying and crusting of the soil, which are the sub-types of degradation most frequently observed in the area. These results are confirmed by authors such as A. SIDO YACOUBA et al, (2018) who, through their investigations carried out in Guillé Koira (Tondikandia) deduced that the soil texture in this locality is dominated by sandy-loam fractions. The same observation was made by M. BAHARI IBRAHIM, (2009) who shows that the granulometry of his study area is dominated by fine sands with very low silt and clay content on all the transects carried out, hence their vulnerability to the risk of erosion. These same soil types were also observed at Bogodjotou and Gassikaïna (in western Niger) by I. BOUZOU MOUSSA and N. DAN LAMSO in 2004. DAN LAMSO in 2004. They confirmed that these soils are poor in organic matter and rich in silts and fine sands. This facilitates the rapid formation of superficial films or surface crusts, especially as they are always poorly protected against the impact of raindrops.

4.1.2. Land use and land degradation

Processing of satellite images shows that the land in this locality underwent a profound change in its occupation and use between 1990 and 2020. This change can be seen in the progression, regression and even disappearance of certain units. Indeed, the extension of cultivated areas and the reduction in fallow land can be linked to demographic growth, as evidenced by human settlements, and to unpredictable climatic conditions that no longer meet the population's expectations. This encourages farmers to increase the size of their fields while abandoning the practice of fallowing. The disappearance of regular tiger scrubland in favour of degraded scrubland may be due not only to human action through deforestation, but also to climatic droughts, which have led to an enormous loss of vegetation on plateaux and slopes, resulting in the spread of rocky terrain. As for the creation of water bodies, this can be justified by the denudation of the plateaux, which can no longer retain the water that has fallen. This water then runs off to return to low-lying areas, hence the appearance of several bodies of water in recent years. These results are in line with those obtained by researchers such as I. BOUZOU MOUSSA et al. 2009, A. INGATAN WARZAGAN, 2013, S. TAHIROU, 2017 and B. OUEDRAGO et al.2019 who revealed that the Sahel is facing changes in its natural landscape linked to climate change and land use.This situation has consequences for natural resources and the services they provide. According to the results of this same analysis, the land in the zone is slightly to very severely degraded. Moderately degraded land accounts for most of the area, with 46.26%. This is found only in the cultivated areas, which are more or less tilled by the farmers. This class is followed by those with severe and very severe land degradation, whose areas are evaluated at 24.09 and 18.09%, or more than 1/3 of the total area of the terroir. These degradation indices are observed on the units of degraded and regular riparian formations, bodies of water and some cultivated areas located on the upstream glacis for heavy degradation and on degraded scrubland, degraded tiger scrubland, bare surfaces and rocky terrain for severely or very severely degraded land. The slightly degraded proportion is the class in last place, with an estimated area of 11.56%. This low level of degradation is observed in the fallow units and a few crop fields around the villages, which are less affected by erosion. These results are similar to those obtained by S. TAHIROU (2017) who adopted the same methodology to assess land degradation in the commune of Sokorbé. Their study shows that in this commune, 3.56% of the land is slightly degraded, 49.12% is moderately degraded and the areas that are severely and very severely degraded are 40.25% and 7.07% respectively.This slight difference between the results can be explained by the fact that the locality in which this study was carried out (Fandou Mayaki) is a very sandy area located on one of the arms of the dallol bosso, where the effects of erosion are much less attenuated than in the other environment. Rain-fed agriculture is the most developed activity in this area. The practice of this activity has led to changes in land use and occupation in this locality. Bare soil and encrusted surfaces can be seen everywhere, reflecting the degraded state of the environment. In addition, plateaux that were once covered by vegetation are now devoid of such cover, giving way to rocky terrain. This situation seems to be general throughout western Niger, as shown by several studies (LOIREAU, 1998; S. YARO, 2008; I. BOUZOU MOUSSA. et al. BOUZOU MOUSSA et al (2011); T. ABDOURHAMANE, (2011) and ABBA (2013) quoted by S. TAHIROU et al (2019). In addition, this degradation phenomenon is manifested in this area by the undermining of the banks of the koris, the silting up and silting up of bodies of water, and the formation of nebkas at the foot of trees or other obstacles. These results concur with those of B. BAGAGI ARZIKA, (2017) who, according to his analyses, revealed that the various forms of degradation recorded in the Makéra II (Matankari) water catchment area include deflations, gullies, dune formation, undermining of banks, silting, encrustation and reduction in plant cover. But there is a difference between its degradation index and what we obtained. In its catchment area, strongly to very strongly degraded units occupy 87.24% of the catchment area, compared with 11.08% for moderately degraded units and 1.68% for slightly degraded units. This phenomenon of degradation will certainly have repercussions on ecosystem services.

4.1.3. Perceptions of farmers on the supply of ecosystem services

Every human being in the world is wholly or partially dependent on ecosystems and the services they provide, such as food, water, disease treatment, climate regulation, spiritual fulfilment and recreational enjoyment (MEA, 2005). However, the results of data collected from the population have shown that these ecosystem services are no longer fulfilling their function. Supply services are being disrupted both by changes in practices and by the disappearance of these practices. This is the case, for example, with agriculture, the population's main activity, which has changed from rain-fed farming in the 90s to rain-fed and market gardening in the last ten years, with a diversification of crop varieties. This is due to land degradation, which compromises rain-fed agriculture, and to make up for the agricultural deficit, the population has decided to practise market gardening as a complementary means. This idea is in line with what S. TAHIROU in 2017 in the commune of Sokorbé, where 64.10% of households surveyed reported that land degradation had led to a drop in their agricultural production over the past five (5) years and was hampering the satisfaction of food needs. This situation is also confirmed by table 6, which shows that agricultural production has shown little growth in recent years. Livestock farming has also suffered the same fate as agriculture. There have also been fluctuations in this activity. The types of livestock farming identified in the zone are sedentary, which is practised by more than 98% throughout the period from 1990 to 2020, while transhumant and sedentary livestock farming are practised by 37% in 1990 and 6% in 2020. Farmed species are also affected by this situation, as some have completely disappeared from the area. The most important livestock species are goats and sheep. This is due to the fall in fodder production, which is one of the consequences of climate change and land degradation, and to the fact that these species are the easiest to rear.

J. BLAMAT (2004) also made this point. He states that in Burkina Faso the sustainability and survival of livestock farming is threatened by the combined effect of agronomic, demographic and climatic factors, characterised by the increasing scarcity of fodder resources, which form the basis of livestock feed. Other supply service products are also facing the same phenomenon of destabilisation under the amplifying effect of climate change, as well as the imbalances between man and nature. Regulating services, for their part, seem to be the most disrupted in this dynamic environmental context characterised by climate variability or change, loss of biodiversity and advanced land degradation.Under these conditions, it is obvious that the regulation between the various biophysical elements is disturbed. In fact, the loss of plant cover, resulting in a rise in temperature, the increase in water and wind erosion, and the disruption of the roles played by micro-organisms are major problems that hamper regulation services. As a result of this situation, the hydrological cycle, which transfers water between the atmosphere, rivers and groundwater, is failing, as it no longer plays its role of sharing rainfall between infiltration and run-off and its capacity to store water for plants and soil organisms. It also leads to a significant loss of carbon and considerable greenhouse gas emissions, which in turn contribute to climate change. This is reflected in the emergence of climate-related risks that existing plant formations are no longer able to mitigate, as shown in Figure 14. All these parameters show that the provision of regulation services is also compromised. This has a negative impact on the self-maintenance service. This can be seen from Figures 11 and 12 which show that in the study area, areas reserved for animals and plants are being nibbled away by farmers, leading to their shortening or even disappearance. This issue of ecosystem services has been addressed by several researchers, who have come up with similar results. This is the case of the 2005 Millennium Ecosystem Assessment report, which estimates that around 60% (15 out of 24) of the ecosystem services studied as part of the Millennium Ecosystem Assessment are being degraded or exploited in an irrational way. In addition to this report, L. Bruno et al (2015) also made this observation, stating that most biological communities are disappearing under the combined effect of climate change and other anthropogenic pressures, affecting the functionality of ecosystems close to collapse.

On the basis of all these analyses, which are in line with what has been revealed in the study area, there is no doubt that the effects of climate change, in correlation with human pressures, are having a major impact on the natural resources on which the lives of many of the world's populations depend. This is disrupting their survival and their very existence. That's why we need to propose development alternatives to mitigate this phenomenon. which hinders the socio-economic development of the region.

4.2. PROPOSED DEVELOPMENT ALTERNATIVES

In view of the worrying situation of land degradation and ecosystem services observed in this area, it is necessary to adopt alternative development solutions. In a context marked by climatic variations and overexploitation of resources, leading to water deficit, loss of vegetation cover and soil loss, restoration is of paramount importance and is becoming a necessity to increase land productivity and restore ecosystem functions and services. As a result, a wide range of rainwater harvesting and land management practices have developed over recent decades to address the negative impacts of land degradation and increase long-term agricultural productivity. To this end, without being too exhaustive, the following techniques and practices are proposed in order to achieve an agroecological system that enhances multiple ecosystem services and increases overall crop productivity

4.2.1. Water and soil conservation techniques

To conserve rainwater and preserve the soil, a number of practices have been developed over the last few years. These practices capture and retain rainwater before it runs off agricultural fields, and also help to trap the sediments that fertilise the soil. They include zaï or tassa, stone cordons and half-moons.

4.2.1.1. Zaï or tassa

Zaï or tassa is an ancient farming technique rediscovered after the great drought of 1973/74 and then perfected by the various people working with farmers. It involves making seed holes about 30 to 40 cm in diameter and 10 to 15 cm deep. The distance between holes is 70 to 80 cm, giving around 10,000 holes per hectare. These holes are dug perpendicular to the slope and in a staggered pattern. This staggered arrangement of holes ensures optimum collection of run-off water and slows the flow of water over the land. The zaï technique concentrates and conserves nutrients and water close to the roots of cultivated plants. For most farmers, applying organic fertiliser directly to the plants is an economical use of a limited production factor. Applying organic manure to the holes helps to restore biological activity, improve fertility and loosen the soil. Zaï are dug in marginal or degraded land that is no longer cultivated, such as low-slope glacis and crusted land in areas with rainfall of less than 800 mm per year. On the other hand, it is not recommended on sandy soils and in low-lying areas. On sandy soils, the holes are not stable. In low-lying areas, the zaï risk being flooded. The zaï technique is particularly interesting in areas where land pressure is high. It can also be used to reclaim additional land. Photo 9 is a perfect illustration of the effectiveness of the zaï technique for restoring degraded land.

Photo n°9 : Zaï sown in a crop field, **Source** : Internet

4.2.1.2. The stone cordons

The stone cordons are anti-erosion devices made up of blocks of rubble/pebbles assembled in series of two or three. They are built in lines along a contour perpendicular to the slope after stripping 10 to 15 cm of soil along the line, as shown in photo 10. The tops of the stones reach a height of 20-30 cm from the ground. The distance between the stone lines is 20 to 50 m, depending on the slope of the land (GIZ, 2012).

They can be found on plateau summits, slopes and steeply sloping glacis where the soils are sandy-silty to silty-sandy, indurated or encrusted. Stone barriers are a measure to combat water table erosion caused by run-off. They slow down run-off and distribute it more evenly over the landscaped area. Reducing the speed of run-off encourages water infiltration, thereby preventing the loss of rainwater. By stopping the water, the cordons encourage the sedimentation of fine soil particles carried along by the water and manure.

Surplus water infiltrates into the soil and through the cordons. In the event of irregular rainfall, the stone cordons thus help to retain more moisture in the soil over a longer period and reduce water shortages during pockets of drought. These techniques work best when combined with biological measures (grassing, hedgerows, tree planting), organic fertilisers and mulching.

Photo n°10 : Landscaped glacis with stone cordons, **Source** : Internet

4.2.1.3. Half moons

The half-moon is a compacted earth or stone structure in the shape of a semicircle with openings perpendicular to the direction of water flow and arranged in a staggered pattern. This staggered arrangement of half-moons perpendicular to the direction of surface water flow means that run-off water can be captured and made available to plants. In this way, the half-moons reduce the loss of water and fertile soil layers. In the medium term, the half-moons encourage sedimentation and contribute to the recovery and protection of bare and encrusted land for agricultural, pastoral or forestry purposes. In this way, the land inside the half-moons, enriched with organic fertiliser, is used for growing cereals (agricultural half-moons), planting woody species and/or sowing with herbaceous plants (sylvopastoral half-moons). Like other anti-erosion measures, half-moons reduce the speed of water run-off and help to improve the quality of the water. This is particularly beneficial when rainfall is low and the half-moons direct water towards the plants, thereby increasing water availability.

In the case of agricultural half-moons, crops survive temporary periods of drought. This technique is used to recover earthworks, especially on plateaux, but it can also be used on lower slopes and steeply sloping glacis. Photo 11 shows a half-moon site developed by ELD (2019) in the rural commune of Badaguichiri (Tahoua).

Photo 11: Site developed in half-moons, **Source:** A. ABDOU et al, ELD Tahoua, (2019)

This technique has made it possible to recover degraded land in this area. In addition, it has also enabled beneficiaries to obtain more diversified and more important ecosystem services in terms of quantity, (A. ABDOU et al. 2019).

4.2.2. Techniques for managing land and ecosystem services

In an effort to respond to the negative impacts of land degradation and ecosystem services, land management practices have been inventoried in recent decades.in order to optimise the management of natural resources and increase agricultural productivity. In the Sahelian region, the most favourable practices are agroforestry, conservation agriculture and integrated soil fertility management.

4.2.2.1. Agroforestry

According to R. WINTERBOTTOM et al (2013), agroforestry is a land use practice that consists of deliberately integrating perennial woody trees on the same plots as crops or livestock. Trees and shrubs in agroforestry systems can be regenerated and protected, or selectively planted and managed. Agroforestry systems can include both native and introduced non-native species. Where farmers include woody species that produce timber, fodder, edible leaves and other products, agroforestry systems evolve into more complex production systems that can provide a wider range of benefits, and more resilient farming systems than those based on annual cereal production alone. These perennial woody trees used in agroforestry interact with soils and crops to create an agro-ecological system that enhances multiple ecosystem services and increases overall crop productivity.

Photo 12 shows an overview of agroforestry.

Photo n°12 : Association of crops and parks with accacia albida, **Source :** Internet

4.2.2.2. Conservation agriculture

Conservation agriculture, as shown in photo 13, is a practice aimed at improving soil fertility and water use efficiency, especially on large farms. It is based on three related practices

Minimal soil disturbance through reduced tillage or the adoption of low-tillage techniques.

suitable for cultivation ;
-retention of crop residues or maintenance of a cover crop ;
-crop rotation or diversification of species grown in sequence and/or in association.
Combining these techniques boosts the soil's organic matter content, retains soil moisture and protects arable land from erosion.

Adopting this practice increases rainfall infiltration, reduces runoff and pollution, improves micro-organisms and sequesters soil carbon, as well as saving on labour costs R. WINTERBOTTOM et al. (2013).

Photo n°13 : Crop residues stored in a field until the next semi, **Source :** Internet

4.2.2.3. Integrated soil fertility management

Integrated soil fertility management is another land and water management practice. to protect resources from degradation. In its broadest sense, it refers to optimising the use of soil nutrient stocks, locally available resources and fertilisers to improve land productivity while maintaining or increasing soil fertility. Its basic practices focus on the combined use of judicious but sufficient quantities of fertiliser and available soil conditioners. These soil improvers come mainly from local sources of organic matter, but may also include lime and phosphate rock. These land management practices make it possible to maintain agro-ecological balance, ecological resilience and the stability of ecosystem services, while ensuring subsistence and a variety of livelihoods for humans (R. WINTERBOTTOM et al. 2013).

CONCLUSION OF THE CHAPTER

Discussion of the results in comparison with those found by other researchers shows that the study area shares almost similar characteristics in terms of land degradation and ecosystem services with all the terroirs that have been the subject of similar studies. To mitigate this As a result, development alternatives have been proposed to improve the quality of these endangered resources.

GENERAL CONCLUSION

Like many other research projects, this study is also part of a continuing effort to contribute to a deeper understanding of the degradation of natural resources and the importance of corrective measures. Its aim is firstly to assess the degradation of land and ecosystem services and secondly to identify management methods that can effectively combat land degradation and consequently preserve ecosystem services. This research was made possible through field observation and the use of satellite images, on the one hand, and the analysis and interpretation of climatic data (rainfall), granulometric data and data from surveys of the population, on the other. To this end, the following results are obtained on the basis of the hypotheses.Observation and use of satellite images have revealed that the land in the Fandou area has undergone a number of changes through land use and occupation. These environmental changes, produced by man interacting with nature, have led to the degradation of natural resources. This is demonstrated by the degradation index map, which shows that 46.26% of the land has been moderately degraded, with heavily and very heavily degraded land accounting for more than 42%. In addition, the climatic data showed that the climate is subject to strong and increasingly recurrent variations, resulting in alternating wet and dry phases. These different phases have had negative repercussions on the environment, modifying its functioning either by disrupting its ecological balance or by destabilising the surface conditions of the soil. Secondly, the analysis and interpretation of the survey data provided the population's opinion on the provision of ecosystem services and land management methods. They perceive that the services provided by the ecosystem are also being disrupted by man or nature. This reduces their capacity to supply goods to the population, which has also seen a huge decline in the services provided by most ecosystem components. In addition, the same data showed that the management methods adopted by the population of this terroir are essentially based on the use of organic manure, mulch, household waste, etc., as well as on the use of pesticides. Finally, granulometric data was used to characterise the texture of the soils in the area. It was found that the soils in this locality are mostly composed of sand. This means that they are both sensitive to erosion and vulnerable to other types of degradation, thus compromising their agro-sylvopastoral production capacity.

These results confirm our research hypotheses that :

- The degradation of land and ecosystem services is a major concern in this area.
- Land degradation factors are due to climatic variability, nature, occupation, etc. the way the land is farmed and managed in this area.
- Possible management alternatives for preserving land and ecosystem services This means adopting good water and soil management and conservation practices.

For the purposes of future research, it would be important to continue these investigations on a local scale and try to assess the impact of this phenomenon on the socio-economic life of the population.

BIBLIOGRAPHICAL REFERENCES

ABDOU Amani, ABDOULAHI SOULEY Chipkao, TIDJANI MAMAN Amina, HASSIMI MOUSSA Seyni, SABRA ABDOUL KADRI Amadou, SOULÉ AHMADOU Elhadji , KAIRE Maguette (2019). Economics of land degradation in Tahoua, Niger. Cost-benefit analysis of land reclamation activities (benches, half-moons and stone cordons) at four sites in the rural commune of Badaguichiri. A report by the ELD Initiative as part of the project "Reversing land degradation in Africa through the large-scale adoption ofagroforestry". 44 p. Available at www.eld- initiative.org.

ABDOULAYE Dia, AMADOU MOCTAR Niang (2010). "Le Projet Majeur Grande Muraille Verte de l'Afrique : contexte, historique, approche stratégique, impacts attendus et gouvernance". In: Le projet majeur africain de la Grande Muraille Verte Concepts et mise en œuvre. IRD: 11-27.

ADDAM KIARI Saidou, ABOUBACAR Ichaou (2016). "Sustainable soil management in Niger: Constraints, challenges, opportunities and priorities" In: La Revue Nature & Faune, Volume 30, Number 1: 31- 34.

AMADOU Ousmane (2016). Behaviour of dug structures according to soil texture of morpho-pedological units in the Kobi sub-catchment (Commune of tondikandia). Master's thesis, Faculty of Agronomy, Abdou Moumouni University of Niamey. 52 p.

AMONMIDE I, DAGBENONBAKIN G, AGBANGBA C.E and AKPONIKPE P. (2019). "Contribution.à l'évaluation du niveau de fertilité des sols dans les systèmes de culture à base du coton au Bénin". Int.J. Biol. Chem. Sci. 13(3) :1846-1860.

AUDRIN Michaëla (2005). Erosion in the Sahel: A study of soil infiltrability in a catchment area Niger. Final dissertation. ISTOM, School of International Agro-Development. P.76.

BAGAGI ARZIKA Binta (2017). Vulnerability of surface water resources to climate change and human activities: the case of the Makera ii waterhole in the rural commune of Matankari. Master's thesis, Department of Geography, FLSH/UAM. 55 p.

BAHARI IBRAHIM Mahamadou. (2009). Dynamique hydrogéomorphologique actuelle du Kori Mountséka (Département de Birni Konni, Région de Tahoua) et ses effets socioéconomiques : approche méthodologique. DEA dissertation, Department of Geography, Faculty of Arts and Humanities, Abdou Moumouni University. 65 p.

BARBANT Pierre (2008). Human activities and land degradation Indicators and indices. IRD.365 p.

BARBANT Pierre (2010). A method for assessing and mapping land degradation. Proposal for standardised guidelines. Les dossiers thématiques du CSFD. N°8. August 2010. CSFD/Agropolis International, Montpellier, France. 52 p.

BLAMAH Jalloh (2004). Impact des aménagements antiérosifs sur les productions fourragères et le recyclage des nutriments au sahel. Final dissertation, Institut du Développement Rural de l'Université Polytechnique de Bobo-Dioulasso. 97 p.

BOUBACAR NA-ALLAH Abdoulaye (2014). Influence des aménagements antiérosifs sur la régénération du couvert végétal et la récupération des sols encroûtés dans les bassins versants de Tondi Kiboro (Sud-Ouest du Niger). Master's thesis, Department of Geography, Faculty of Letters and Human Sciences, Abdou Moumouni University.58 p.

BOUZOU MOUSSA Ibrahim, DAN LAMSO Nomaou (2004). "Le " Tassa " : une technique de conservation des eaux et des sols bien adaptée aux conditions physiques et socio-économiques des glacis des régions semi-arides (Niger) / The "tassa", a soil and water conservation technique well adapted to the physical and socio-economic conditions of the glacis of semi-arid regions (Niger)". In: Revue de géographie alpine, Tome 92 N°1 : 61-70.

BOUZOU MOUSSA Ibrahim, OUMAROU FARAN Maiga, JEAN-MARIE KARIMOU Ambouta,
BENOIT Sarr, LUC Descroix, MAHAMAN MOUSTAPHA Adamou. (2009). "Les conséquences géomorphologiques de l'occupation du sol et des changements climatiques dans un bassin-versant rural sahélien". Research article, Sécheresse;20(1):145-152.

BRUNO Locatelli, MIGUEL Pedrono, DRISS Ezzine, DENIS Pesche, SERGE Morand, AURÉLIE Binot (2015). "Ecosystem services in the face of climate change" in Academia Accelerating the world's research :236-245.

CHEBBANI Rachid, DJILLI Kaddour, E. ROOSEric (1996). "Étude à différentes échelles des risques
d'érosion dans le bassin versant de l'isser, Algérie". Bull. Erosion n°16. : 85-95

CHRISTIAN Walter, ANTONIO Bispo, CLAIRE Chenu, ALEXANDRA Langlais and CHRISTOPHE Schwartz (2015). "Soil ecosystem services: from concept to valuation". Cahiers Demeter: 53-68.

UNITED NATIONS CONVENTION TO COMBAT DESERTIFICATION(CNULD). (2015). Climate change and land degradation: Linking knowledge to issues, results of the 3rd UNCCD Scientific Conference 9-12 March, Cancún, Mexico. 19 p.

DAN GOMMA Abdou, ISSA Salissou, ABASSE Tidjani, BANOIN Maxim, SCHLECHT Eva (2019). "Effects of fodder collection on the environment and on agro-sylvo-pastoral and commercial activities in Sahelian regions in Niger". Int. J. Biol. Chem. Sci. 13(6): 2590-2602.

EDWIGE Botoni, CHRIS Reij (2009). The silent transformation of the environment and production systems in the Sahel: Impacts of public and private investment in natural resource management. Summary report of the SAHE study on Burkina Faso, Mali, Niger and Senegal. 63 p.

ELD Initiative. (2019). ELD Campus. Module: Land degradation versus sustainable land management. 32 p. Available at www.eld-initiative.org.

ELHADJI MOUSSA Sani (2007). Évaluation du ruissellement et de l'érosion potentiels à Wankama, degré carré de Niamey. Master's thesis, Geography Department, Faculty of Letters and Humanities, Abdou Moumouni University. 60 p.

FOLK R L, WARD W C. (1957). "Brazos rivers bar: a study in the significance of grain size parameters". Journ. Sc. Petrol. Tulsa Okla, N°1, Vol. 27: 3 - 26.

FOURATI Marwa, BOUAZIZ Riadh, ELAMRI Asma, MAJDOUB Rajouene (2015). "Water and soil conservation techniques in the oued sidi Salah catchment (sfax-tunisia): evaluation trial". Algerian journal of arid environment 26 vol. 5, no. 2, December 2015: 26-38.

GEMENNE François, BLOCHER Julia, DE LONGUEVILLE Florence, VIGIL DIAZ TELENTI Sara,
ZICKGRAF Caroline, GHARBAOUI Dalila and OZER Pierre (2017). Climate change, natural disasters and population displacement in West Africa. 22 p.

IPCC. (2013). Glossary [Planton, S. (coord.)]. In: Climate Change 2013: The Scientific Evidence. Contribution of Working Group I to the Fifth Assessment Report of the Intergovernmental Panel on Climate Change, Cambridge University Press, Cambridge, UK and New York, NY, USA: 185-204.

GIZ. (2012). Good water and soil conservation practices, a contribution to climate change adaptation and producer resilience in the Sahel. German Federal Ministry for Economic Cooperation and Development. 59 p.

YADJI Guéro, N. DAN LAMSOomaou (2006). Natural resource restoration projects and soil fertility. Sahelian Studies, CRESA DE NIAMEY. 34 p.

HASSIMIOU HALIDOU Djabri, ABDOU MAMAN Manssour, ALZOUMA MAYAKI Zoubeirou (2020). "Characterisation of the soil of the degraded site of Sakey koira Tegui in Niger for better reforestation with Acacia senegal". Int. J. Biol. Chem. Sci. 14(4): 1470-1478.

ILBOUDO Adama, SOULAMA Soungalo, HIEN Edmond, ZOMBRE Prosper (2020). "Farmers' perceptions of the degradation of natural resources in lowlands in the Sudano-Sahelian zone: the case of the Nakanbé-Dem sub-watershed in Burkina Faso". Int. J. Biol. Chem. Sci. 14(3): 883-895.

INGATAN WARZAGAN Agahli (2013). Aménagements antiérosifs et écoulements dans un petit bassin versant expérimental endoréique (sud-ouest du Niger). Master's thesis, Department of Geography, Faculty of Letters and Humanities, Abdou Moumouni University. 93 p.

INGATAN WARZAGAN Aghali (2011). Consequences of the degradation of the tiger bush of Fakara (southwest Niger) on the genesis and increase of runoff. Master's thesis,

Department of Geography, Faculty of Letters and Humanities, Abdou Moumouni University. 82 p.

INSTITUT NATIONAL DE LA STATISTIQUE (2015). State and structure of the population of Niger in 2012. 54 p.

JONATHAN Lash, BJÖRN Stigso, JOHN Ehrmann (2009). Valuation of Ecosystem Services to Business: A Practical Guide to Identifying Risks and Opportunities from Ecosystem Change, Version 1.0. 43 p.

KADER Mohamed, HAMIDOU Djibo, PHILIPPE Morant (2008). Restoring degraded land in Niger's river valley: a successful experiment in participatory management of fragile natural resources. APPRI2008 International Workshop. Oct 2008, Ouagadougou, Burkina Faso. CIRAD-00406261. 4 p.

KOHIO Edmond, TOURE Alassane, SEDOGO Michel, AMBOUTA Karimou. (2017). "Constraints to the adoption of good sustainable land management practices in the Sudanian and Sudan-Sahelian zones of Burkina Faso". Int. J. Biol. Chem. Sci. 11(6): 2982-2989.

KOUSSA M, BOUZIANE T. (2018). "Contribution of GIS to the mapping of areas at risk of erosion.
hydrique dans la région de Djelfa, Algérie". Journal Scientifique Libanais, 19(1) :31-46.

LAMINOU MANZO Ousmane, AMANI Abdou, DAN GUIMBO Iro, RACHIDI ABDOU Harouna,
MAHAMANE Ali (2020). "Impacts of banquettes in the recovery of degraded land in Niger". in Journal of Applied Biosciences 151 :15510 - 15529.

MAHAMANE Ali, SALEY Karim, SAADOU Mahamane, JAUFFRET Sandrine, WATA Issoufou, YACOUBOU Bakasso, MOROU Boubé, DIOUF Abdoulaye, MAMANE MAAROUHI Inoussa, ISSAKA Abassa, ZAMAN-ALLAH Mainassara, ICHAOU Aboubacar, ABOUBACAR Issa, TANIMOUNE Arzika. (2009). Climate variability in Niger: potential impacts on vegetation distribution. 14th International Colloquium on Environmental Assessment - Niamey on the theme: Climate change and environmental assessment - issues and tools for impact assessment and the development of adaptation plans. 28 p.

MALAM ABDOU Moussa. (2005). Genèse et morphodynamique actuelle des bas-fonds sahéliens : étude comparative de quelques bas-fonds ruraux de socle et de bassin sédimentaire dans l'ouest nigérien. Master's thesis, Department of Geography, FLSH/UAM. 141 p.

MEA. (2005). Millennium Ecosystem Assessment Synthesis Report Final draft for review and printing. 159 p.

MINISTERE DU PLAN. (2020). Second voluntary national report on sustainable development goals in Niger. 111p.

MOUKAILA Yayé (1988). Pastoral use of the plateaux of the Centre-East unit of the Canton of Tondikandia. Rapport de stage. Faculty of Agronomy, Abdou Moumouni University, Niamey. 42 p.

OUEDRAGO Blaise, KABORE Oumar, KABORE Martine (2019). "Quantitative mapping of soil erosion by GIS/ RUSLE aproach in the Commune of Karangasso Vigué, Burkina Faso". Int. J.Biol.Chem.Sci 13(3): 1638-1653.

OUSSEINI SIDIKOU, Fati (2015). Effets des ouvrages CES/DES sur la résilience des socio-économiques de la zone d'intervention du projet REGIS-ER : cas de la commune rurale de Tondikandia (Département de Filingué, région de Tillabéry, Master's thesis, Department of Geography, FLSH/UAM. 55 p.

OZER Pierre, DJABY Bakary, DE LONGUEVILLE Florence. (2015). Évolution récente des pluviométriques au Niger (1950-2014), Workshop "Coopération Universitaire au Développement avec le Niger" Université de Liège, Arlon Campus Environnement. 18 p.

PATUREL Jean-Emmanuel, SERVAT Eric, DELETTRE Marie-Odile, LUBES-NIEL Hélène (1998). Analysis of long-term rainfall series in non-Sahelian West and Central Africa in a context of climatic variability. Journal des sciences hydrologiques 43(6). 937-946

COMMUNAL DEVELOPMENT PLAN (PDC). (2020). Plan du développement communal réplanifié de la commune rurale de Tondikandia. 184 p.

RESCCUE PROJECT (2016). Assessment of the loss of ecosystem services associated with mining activity.
V2. 63 p.

REPUBLIQUE DU NIGER, MINISTERE DE L'ENVIRONNEMENT ET DE LA LUTTE CONTRE LA DESERTIFICATION (2001). Nomenclature pour la construction de bases de données sur l'occupation des sols au Niger au sud du 16eme parallèle, final version. Direction de l'environnement. 70 p.

RIVES Fanny, DENIS Pesche, PHILIPPE Méral, STÉPHANIE Carrière. (2016). Ecosystem services: a debated concept in ecology. 22 P.

ROOSE Eric, DE NONI George (1998). "Apport de la recherche à la lutte antiérosive, bilan mitigé et nouvelle approche". ORSTOM, Laboratoire de comportement des Sols cultivés, Montpelier. Études et gestion des Sols .5.3: 181-194.

SEYNI MOUSSA Ibrahim (2008). État des lieux des zones aménagées par le Projet Agro-Sylvo-Pastoral (P.A.S.P) Tillabéri-nord et essai d'élaboration d'un Système d'Information Géographique (SIG) pour un suivi environnemental : cas de la commune de Kourthey (Sansané Haoussa). DEA dissertation, Geography Department, FLSH/UAM. 93 p.

SIDO YACCOUBA Amir, YADJI Guero, MAMANE TCHIKAMA Mella, ABDOURHAME ISSA
Nourou (2018). "Characterization of Guillé Koira and Lassourou soils in the rural communes

of Imanan and Tondikandia in Niger for improved rice cultivation around ponds". Int. J.Biol. chem. Sci.12(6): 2474-2485.

TAHIROU Souley, WAZIRI MATO Maman, ABBA Bachir (2019). " Dégradation des terres et vulnérabilité des populations de la commune rurale de Sokorbé face au changement climatique et a la pression anthropique ". Revue de Géographie de l'Université de Ouagadougou, No. 08, Volume 1, October 2019: 19-41.

TAHIROU Souley. (2017). Land degradation and vulnerability of populations in the rural commune of Sokorbé (Dosso region) in a context of climate change and anthropogenic pressure. Master's thesis, Department of Geography, FLSH/ UAM.79 p.

TOSSOU EDJROGAN Michaël, ADECHINA ARCHANGE Rodrigue, HOUNKPE BIDOSSÈSSI Jéchonias, KELOME CARINE Nelly, SY BOUBOU Aldiouma, DIAW AMADOU Tahirou (2019)."Granulometric Analysis and Mapping of Superficial Sediments of the Benin Coastal Lagoon (West Africa)". European Scientific Journal, Vol 15, N°24, Doi:10.19044/esj. 2019.v15n24: 294- 314.za

TRASK P. (1932). Origin and environment of source sediments of petroleum Golf publishing Co, Houston, Texas, U.S.A. 323 p.

WENTWORTH K H. (1922). "A scale of grade and class terms for clastic sediments". Journal of Geology 30: 377-392.

WINTERBOTTOM Robert, CHRIS Reij, DENNIS Garrity, JERRY Glover, DEBBIE Hellums, MIKE Mcgahuey, SARA Scherr (2013). Improving land and water management. Discussion paper, episode 4 of Creating a Sustainable Food Future. Washington, DC:WorldResourcesInstitute. 44 P. Available at: http://www.worldresourcesreport.org.

YAMBA Boubacar (1989). "Land pressure and degradation of the natural environment". In: aménagement rural en Afrique de l'Ouest: le cas du Niger. Université Libre de Lausanne: 29-44.

MIX
Papier aus verantwortungsvollen Quellen
Paper from responsible sources
FSC® C105338

Printed by Books on Demand GmbH, Norderstedt / Germany